REACTIONS WITH VARIABLE-CHARGE SOILS

The soil was for a long time regarded as a collection
of insoluble inert mineral fragments admixed with
small quantities of more soluble substances . . .
This view had the advantage of simplicity . . .
But unfortunately it leads to inaccurate conclusions

Russell & Prescott, Journal of Agricultural Science 1916

Developments in Plant and Soil Sciences

Reactions with Variable-Charge Soils

By

N.J. BARROW
CSIRO, Division of Animal Production
Wembley, Western Australia

1987 **MARTINUS NIJHOFF PUBLISHERS**
a member of the KLUWER ACADEMIC PUBLISHERS GROUP
DORDRECHT / BOSTON / LANCASTER

Distributors

for the United States and Canada: Kluwer Academic Publishers, P.O. Box 358, Accord Station, Hingham, MA 02018-0358, USA
for the UK and Ireland: Kluwer Academic Publishers, MTP Press Limited, Falcon House, Queen Square, Lancaster LA1 1RN, UK
for all other countries: Kluwer Academic Publishers Group, Distribution Center, P.O. Box 322, 3300 AH Dordrecht, The Netherlands

Library of Congress Cataloging in Publication Data

```
Barrow, N. J.
   Reactions with variable-charge soils / N.J. Barrow.
      p.   cm. -- (Developments in plant and soil sciences)
   ISBN 9024735890
   1. Variable charge soils--Mathematical models.  2. Variable charge
soils--Data processing.  3. Soil solutions--Mathematical models.
4. Soil solutions--Data processing.  5. Soil absorption and
adsorption--Mathematical models.  6. Soil absorption and adsorption-
-Data processing.  7. Ions--Mathematical models.  8. Ions--Data
processing.   I. Title.  II. Series.
S592.53.B37 1987
631.4'1--dc19                                        87-17785
                                                        CIP
```

ISBN-13: 978-94-010-8141-2   e-ISBN-13: 978-94-009-3667-6
DOI: 10.1007/978-94-009-3667-6

Copyright

TABLE OF CONTENTS

Introduction

This book is about ions, about variable-charge surfaces and about models. It is about ions because most of the substances in which soil scientists are interested occur in the soil solution as ions. This applies to both plant nutrients and pollutants. Thus the reaction between soil and say phosphate does not involve a substance called "phosphate"; it involves phosphate ions. Ions are charged particles. When these charged particles react with charged surfaces, the outcome of the reaction is affected by both the charge on the particles and the charge on the surfaces. Hence we also need to understand the charge on the surfaces. If our understanding is adequate, we should be able to express our ideas precisely — ideally by writing equations. Unfortunately the equations turn out to be quite complex and to interrelate with each other in a complex way. If we want to envisage the effect of varying some of the conditions we have to include the equations in computer programs. Because these programs describe physical systems, they are called models.

A word about the role and function of models is in order. A computer model is no more than a precisely expressed hypothesis. Like all hypotheses, it can be used to make predictions. The predictions from some hypotheses turn out to be better than those of others and we are to prefer the one that makes the best predictions. However, in practice, old ideas are not abandoned readily. We cling to them until the new ideas become familiar. For example, some of the models that have been used to describe ion reaction with surfaces have a pleasing familiarity about them. They seem to involve only small modifications of ideas about reactions of ions in solution. Perhaps that is why they have been enthusiastically adopted — even though they are not as good as alternative models. An important role for computer models is to help us to become familiar with the ideas behind the model. The best way to do this is to play with the model. The word "play" is used quite deliberately because of the relation between playing and exploring. If a model is to be played with, it should not be too frightening and it should be readily accessible. In the jargon of computer users, it should be "user friendly". These considerations have influenced my choice of presentation. One possibility was to try to present giant models that attempt to cater for every possible method of use. This "black box" approach does not involve the user in much understanding and, since the possible uses are many, the model can become unwieldy. Instead I have tried to present programs that fit some common usages and which can be changed by the user to suit other purposes. This is one reason for using BASIC for the programs — it is fairly easy to make changes and to get them running. The main disadvantage of BASIC is that it has several 'dialects' and this can restrict usage. I have therefore used Microsoft-BASIC as this is probably the most common dialect. The programs were initially developed on other machines but were converted and run using an IBM-compatible personal computer equipped with a "Trump card". This card permits very simple editing of programs and very rapid compiling. The programs used here run on it about ten times faster than on an IBM-compatible machine. It therefore overcomes the main disadvantage of BASIC — its slowness. The Trump card was purchased from Sweet Micro Systems, Inc. 50 Freeway Drive, Cranston, RI 02920, USA.

This book is divided into two sections. The first section describes the observations that are to be explained and then outlines the models that may be used to explain them. It moves from the particular to the general — from the behaviour of substances (such as iron oxides), that have been used as models of soil constituents, to the behaviour of soil itself.

This sequence was chosen because I think that models at any level should be based on, and be compatible with, knowledge at a more detailed level. Thus models of soil should be based on knowledge of the behaviour of soil constituents. Of course, models become impossibly complex if every detail of the behaviour of a constituent is included in models of the behaviour of the whole. So, as we move from the particular to the general some abstracting and simplification is needed. The second section is concerned with making the models work. It begins by discussing the problems of solving the sets of simultaneous equations that comprise many models. It continues by discussing the problems of allocating values to parameters to test whether a model can closely describe a set of data. These two problems are similar because both involve repeated calculations using different test values until the best set of values is found. Finally listings are presented for several of the models described in the first section of the book.

This book is mainly directed towards those research workers in soil science who are interested in either the supply of nutrients to plants from the soil, or in the control of pollutants that have reacted with the soil. However it will also be of interest in a wide range of applications in which ions react with surfaces. Although a fair proportion of the book is devoted to computer programs, it is not a book for computer programmers. I am only an amateur programmer and so there will be no flashing lights or blinking messages. I hope this means that the programs will be useful to, and usable by, the other very amateur programmers who want to explore this subject.

Chapter A1

The ionic species present in soil solutions

It is a basic doctrine of this book that neither "phosphate", nor "zinc", nor any other plant nutrient or pollutant reacts with soil. Rather, individual species — mostly ions — react with soil. Most of the readers of this book will be familiar with the species present in solution for some nutrients such as for phosphate and sulfate — but the range of behaviour of the various elements of interest is large. Understanding this behaviour is essential to later arguments and so will be summarized in this chapter.

Dissociation constants

Consider the acid HA which dissociates:

$$HA \rightleftarrows H^+ + A^- \qquad\qquad A1.1$$

For this reaction, the dissociation constant is:

$$K_A = \frac{[H^+][A^-]}{[HA]} \qquad\qquad A1.2$$

Where square brackets indicate concentrations. Inspection of this equation will show that, when $K_A = [H^+]$, the acid is half dissociated. The negative log of the dissociation constant (pK_A) is thus the pH at which the acid is half dissociated.

Equation A1.2 may also be expressed in terms of activities instead of concentrations. This format gives a more-general value. It can be converted to the value appropriate for a given ionic strength as indicated, for example, by Lindsay (1979).

Acids may also be multibasic — that is they may dissociate two or more protons. They may also be involved in polymerisation reactions or reactions with other ligands. Equations to describe some of these reactions are given in Chapter B1 together with computer programs to caculate the proportions of the species present.

Values for the dissocation constants

Many text books and hand-books tabulate values for dissociation constants. Some of these are reliable; others are not. For example, a range of values can be found for the dissociation constants of hydrofluoric acid and for the second dissociation constant of selenious acid. Obviously, some of the published values are wrong! One good source of values for elements of interest to soil scientists is the book by Lindsay (1979). A much larger range of elements is considered by Baes and Mesmer (1976). This book is entitled "The Hydrolysis of Cations", but, even so, it contains useful sections on some anions including borates, silicates, selenites and molybdates. An attractive aspect of the book is that it is more than a tabulation of observed values; it is also an evaluation. The authors are not afraid to reject observations if they consider them unreliable. They also make it clear that accurate values of some of the constants simply do not exist. This is especially so, where the behaviour is complex — involving perhaps several dissociation steps and several

1

polymers. In these cases, the behaviour can sometimes be described almost equally well by several different schemes and so the allocation of constants is difficult. In the subsequent section, which summarises the behaviour of important elements, Baes and Mesmer (1976) has been used as a major source of constants.

The behaviour of some typical anions

Of the ions of interest, the behaviour of sulfate and selenate is probably the simplest. In both cases, the pK_2 is near 2. At the pH of soil, say 4 and above, they are therefore fully dissociated to give SO_4^{2-} and SeO_4^{2-}. The pK of hydrofluoric acid is near 3. In simple systems it is also virtually fully dissociated at soil pH values to give F^-. However fluoride has a strong affinity for aluminium ions (Lindsay 1979) and, at pH values below about 5.5, there may be appreciable concentrations of aluminium in solution. This can mean that the fluoride is largely present as cationic complexes such as AlF^{2+}.

The behaviour of phosphate appears to be more complex in that it is tribasic. However the first dissociation constant is near 2. This has little influence in the behaviour at soil pH values. Similarly the third dissociation constant, at above 12, also has little influence. Thus, in practice, we are mainly concerned with the dissociation of the acid $H_2PO_4^-$ to give HPO_4^{2-} and H^+. The pK for this dissociation is near 7. Hence, over the range of normal soil pH values, we change from mostly $H_2PO_4^-$ at pH 4 to mostly HPO_4^{2-} at pH 8. Over much of this range, the proportion present as HPO_4^{2-} increases ten-fold for each unit increase in pH (Fig.A1.1). This rate of change falls off as the HPO_4^{2-} ion begins to dominate. These changes in the proportions of the phosphate ions with changes in pH are an important component of the chemistry of soil phosphate. Unfortunately they have sometimes been overlooked in studies on the effects of pH on phosphate retention.

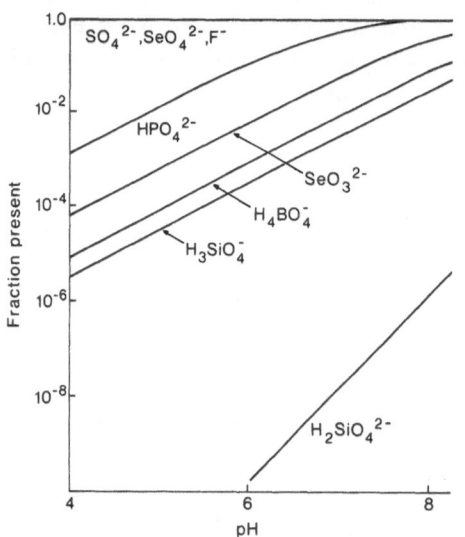

Fig. A1.1 Effect of pH on the fraction present as the indicated species at an ionic strength of 0.03. Thus the line labelled SeO_3^{2-} indicates the proportion of the selenite present as SeO_3^{2-}.

The behaviour of selenite is in some ways, analogous to that of phosphate. The pK for the first dissociation is below 3 and so has little importance for most studies with soil. Like phosphate, we are mainly concerned with the dissociation of the monovalent of $HSeO_4^-$ to give the divalent SeO_3^{2-} and H^+. The pK for this reaction is above 8. Thus the species present in soil solutions are likely to be $HSeO_3^-$ and SeO_3^{2-}. The proportion present as SeO_3^{2-} increases ten-fold for each unit increase in pH (Fig.A1.1).

Boric acid is a weak acid and hydrolyses to give the monovalent $H_4BO_4^-$ ion. The pK for this reaction is near 9. Hence the species present are the uncharged $B(OH)_3$ species and the monovalent ion. The fraction present as the monovalent ion, is low at low pH. However, it also increases ten-fold for each unit increase in pH (Fig.A1.1).

Silicic acid H_4SiO_4 is a very weak acid. Its first pK is near 10 and its second just above 12. Two further dissociations are possible. The pK values for these are thought to be high but no reliable values seem to exist. Polymers can also form but in soil solutions concentrations would not be high enough for this to occur. Thus in soil solutions, the major species is the uncharged acid. However the monovalent $H_3SiO_4^-$, though only a small fraction of the total, increases ten-fold for each unit increase in pH. Furthermore, because the influences of pK_1 and pK_2 overlap, the divalent $H_2SiO_4^{2-}$ increases 100-fold for each unit increase in pH (Fig.A1.1).

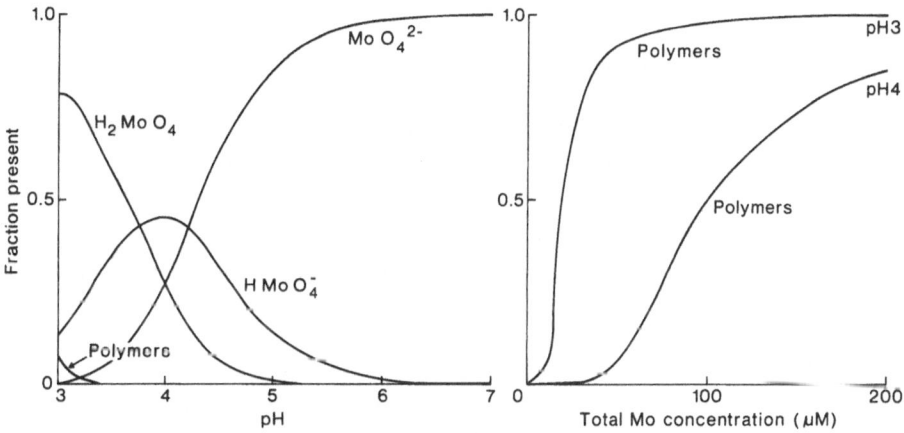

Fig. A1.2 (a) Effect of pH on the distribution of polybdate species at a total molybdenum concentration of 10 μM (0.96 μg Mo/ml); (b) Effect of molybdate concentration on the proportion present as polymers at pH3 and 4.

Molybdic acid is somewhat more complex. The pK_1 and the pK_2 are both near 4. This means that three species are present in the pH range above 4 (Fig. A1.2). The monovalent $HMoO_4$ reaches its maximum concentration near pH 4 and declines at both lower and higher pH values. Another source of complexity is the tendency of molybdates to form polymers. An example of the kind of reaction thought to be involved is:

$$7\ MoO_4^{2-} + 8H^+ \rightleftarrows Mo_7\ O_{24}^{6-} + 4H_2O$$

From the law of mass action, formation of this polymer is proportional to the seventh power of the molybdate concentration and the eighth power of the proton concentration. Thus such polymers become important in acid solutions and with high concentrations of molybdate. The method described in Chapter B1 was used to calculate the effects of pH and of concentration. Fig.A1.2 shows that, at pH 3, polymers are the dominant species present even at low concentrations. At pH 4 they only occur above 20 μM (about 2ppm). At pH 5, the amount present is trivial even at high concentrations.

The behaviour of some cations

Many metal ions are closely associated in solution with a sheath of water molecules. One, or more of these molecules can dissociate a proton. For example:

$$Zn(H_2O)_6^{++} \rightleftharpoons Zn(H_2O)_5\, OH^+ + H^+$$

This reaction is analogous to equation A1.1. It differs only in that the product is positively charged. The process may continue and further protons may be lost. Polymers may also be formed. However, it turns out that the first association product (here, $Zn(H_2O)_5\, OH^+$) seems to be very important in the reaction with soils. The elements of interest may be regarded as forming a sequence with a range of tendencies to lose a proton indicated by the pK_1: Hg^{2+}, 3.4; Pb^{2+}, 7.71; Cu^{2+}, <8; Zn^{2+}, 8.96; Co^{2+}, 9.65; Ni^{2+}, 9.86; Cd^{2+}, 10.08; Mn^{2+}, 10.59. These values are for activities and were obtained from Baes and Mesmer (1976).

With the exception of mercury, the elements have somewhat similar behaviour. The pK_1 is fairly high and so, at soil pH values, the species present are the divalent ion and the monovalent ion. The monovalent ion though present as a small fraction of the total, again increases ten-fold for each unit increase in pH. The behaviour is therefore analogous to some of the anions in Fig.A1.1. Mercury differs in that the pK_1 is much lower. The pK_2 is also low. A graph for the species present therefore resembles that for molybdate (Fig.A1.2) in that the monovalent species (here $HgOH^+$) has its maximum abundance at low pH (about pH 3 Fig.A1.3). At higher pH values, the doubly dissociated species dominate ($Hg(OH)_2^\circ$ and MoO_4^{2-}). Mercury also has a strong tendency to form complexes with halides. This is of special significance for mercury because of its tendency to occur as a pollutant in saline environments. There have therefore been several studies on the effects of chloride on the reaction of mercury with soils or soil constituents. One of the complexes formed is HgOHCl. The abundance of this species depends on both the chloride concentration and the pH (Fig.A1.3). The peak abundance occurs at higher pH values as the chloride concentration increases. At higher levels of chloride, HgOHCl becomes less important as other chloride complexes increase. At moderate levels of chloride it can be a major species.

4

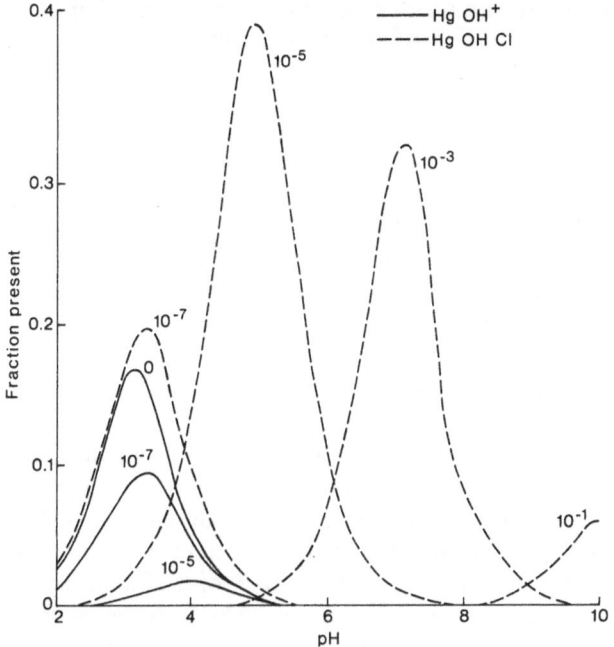

Fig. A1.3   Effect of pH and of the indicated chloride concentrations on the fraction of mercury(II) present as HgOH$^+$ and HgOHCl ions.

References

Baes, C. F. and Mesmer, R. E. 1976. The hydrolysis of cations. John Wiley and Sons. New York.
Lindsay, W. L. 1979. Chemical equilibrium in soils. John Wiley and Sons. New York.

Chapter A2

Variable charge oxides as soil constituents and as models of soil constituents

The oxides present in soils

Many oxides have been identified in soils. For example, the following oxides of iron are known to occur: goethite, $\alpha FeOOH$; lepidocrocite, $\gamma FeOOH$; hematite, αFe_2O_3; maghemite, γFe_2O_3; ferrihydrite, $HFe_5O_8 \cdot 4H_2O$; and magnetite, Fe_3O_4 (Taylor *et al.* 1983). In addition, Taylor *et al.* (1983) list five oxides of aluminium, thirteen oxides of manganese, six of titanium, and four forms of silica. (The term "oxide" is used here in the general sense to include oxyhydroxides).

These oxides can be very important in determining soil properties. Much of the silica may be present as quartz. It is therefore unimportant in soil chemistry — but can have a large effect on soil physics. The colour of soils is often strongly affected by the amount of iron oxide present and by its mineral form. Manganese oxides obviously affect manganese supply to plants but they also have a very strong tendency to react with cobalt. For example, Taylor and McKenzie (1966) found that an average of about 80 per cent of total soil cobalt was associated with manganese oxides but only a small fraction of the total nickel was in this form. The atoms near the surface of the oxide cannot complete the regular pattern of bonds of the atoms in the body of the oxide. They therefore tend to complete their unsatisfied bonds by forming links to water molecules. These water molecules may be thought of as losing or gaining a proton depending on the pH. Thus the charge on the surface varies with pH — that is they are variable-charge materials. The charge so-formed is an important soil property in itself. It is also an important determinant of the amount of reaction with other ions. Iron oxides and aluminium oxides have long been thought to be involved in the reaction of anions with soil. Some of the evidence for this is indirect. It includes correlation studies and studies in which the oxides were either removed from, or added to, soils. However the most convincing evidence comes from direct observations using modern techniques. For example, Norrish and Rosser (1983) used an electron microprobe to show that in many Australian soils much of the phosphorous is associated with iron oxides and little with aluminium oxides. Further, when radioactive arsenate was reacted with soils, it was found that the reaction was largely with the iron oxides (Fordham and Norrish 1979b).

A major problem in studying the chemical reactions of the metal oxides in soils is that it is difficult to remove them from soils and so study them in isolation. This contrasts with the clay minerals which can be removed from soils. This usually involves dissolving the metal oxides. Many of the studies have therefore involved synthesised oxides. While this has yielded much valuable information, the synthesised oxides may differ in several important ways from the material occurring in soils. This is well illustrated in the case of goethite. Goethite crystals are usually produced in the laboratory by aging a ferric hydroxide precipitate at high pH. The crystals so-formed are elongated and they are, supposedly, pure iron oxyhydroxide. This characteristic shape is not seen for soil goethites, and they are far from pure. For soil goethites a surprisingly large proportion of the iron atoms may be replaced by aluminium atoms. The maximum seems to be one aluminium atom per two iron atoms (Schwertmann 1985). There is also some evidence that the higher the aluminium content, the higher the phosphorus content (Norrish and Rosser 1983). (In this case, it seems rather pointless to ask whether the phosphorus has reacted with an iron

6

oxide or an aluminium oxide.) Soil goethites may also contain both phosphate and silicate. An example of a high silica content was given by Fordham *et al.* (1984) for an unusual fibrous goethite from the B horizon of a podzol. For this material the approximate atomic ratios were 4:2:1 for Fe:Al:Si. This is a high content of silicon and corresponds to about five per cent by weight. Nevertheless appreciable contents of silicon are not unusual. Thus Fordham and Norrish (1979a) found that the silicon content of iron oxide pellets for a range of soils ranged between one and four percent by weight. The difficulty in such work is to be sure the sample is free of contamination but these values were for pellets believed to be free of this problem. If we are to understand the behaviour of soil goethite we need to know how these silicon atoms are arranged. Are they, perhaps, scattered at random through the crystal, or are they arranged in some kind of a pattern? This question is difficult to answer for the minute particles of goethite in soil. Fordham *et al.* (1984) pointed out that the silicon apparently occupied a stable position because it was resistant to oxalate extraction. They speculated that the silicon was adsorbed as silicate on the surface of the microcrystals. In this case it would modify the surface charge and might promote aggregation of the microcrystals. It might then form bonds between adjacent microcrystals. This suggestion is consistent with the work of Smith and Eggleton (1983). They examined four samples of botryoidal goethite which varied in silica content from zero to 1.25%. The samples which contained silica were found to be composed of fine, needle-like grains with diameters as small as 30 μM. They suggested that the apparent gaps between these grains could consist of a monolayer of silica. The sample of goethite that did not contain silica did not have this structure.

Thus, in summary, it is possible that soil goethite particles may have a "masonry" structure in which the bricks are composed of (Fe/Al)OOH and the mortar between the bricks is composed of silica. It seems likely that the silica mortar could be partly replaced by phosphate or by other anions.

Synthetic oxides as models of soil oxides

Because of the difficulty of working with oxides extracted from soil, most work has been done with synthetic oxides. The choice has depended on the particular purpose. James *et al.* (1980) pointed out that rutile (TiO_2) has advantages as a model colloid against which to test colloid models. Its advantages are: it is readily available; it is easily purfied; it has a low solubility; the point of zero charge is at a near-neutral pH value; and the particles are of a fairly uniform size and shape. Others have studied oxides which they considered to be important for their particular problems. For example, McKenzie (1980) included the manganese oxide, birnesite, in his studies, because manganese oxides are important in soils for reactions with some heavy metals. However most work has been done with iron oxides because of their great importance. Some workers, for example, Benjamin and Leckie (1981) have used freshly-precipitated iron oxyhydroxide. Others have used well-crystalized materials such as goethite or lepidocrocite. Conditions for the formation of goethite crystals were studied by Atkinson *et al.* (1968). However, even when the same procedures are followed, different batches of goethite can vary appreciably. For example Hingston (1970) prepared batches of goethite which varied in surface area from 17 to 81 $m^2 g^{-1}$. Part of the variation between batches of goethite may be due to differing exposure to carbon dioxide. Evans *et al.* (1979) found that removal of carbon dioxide from goethite was difficult. They passed a stream of wet, carbon dioxide-free nitrogen through a goethite suspension of pH8 for one to two months and performed their titrations of goethite in sealed vessels. Another source of variation is the quality of the water used in experiments with goethite. Evans (1976) reported that deionised water was unsuitable for use in accurate surface chemistry. In his work he used water distilled from alkaline permanganate and he also took precautions to prevent salt and ammonia carry-over into the distillate and to prevent contamination of the water by materials with which it came in contact. After taking these precautions, Evans *et al.* (1979) obtained some unusual results. They found that the point of zero charge of the goethite was at pH 9.3. This is higher than values commonly reported (Kinniburgh and Jackson 1981) but close to a value of pH 9.48 predicted by Yoon *et al.* (1979) from crystallographic data. They also found that there was little pH drift — that is, when acid was added to suspensions of their purified goethite, a stable pH was quickly reached. Others have reported that, under their conditions, the pH values are not stable and changed even over periods of days (Madrid and de Arambarri 1978). These results raise the suspicion that some of the published work with goethite has not, in fact, been done with pure goethite but rather with goethite contaminated to various extents and especially with carbonate.

Another possibility is that samples of goethite prepared by different workers have differed in their crystallinity. Schwertmann *et al.* (1985) showed that, by using a range of temperatures, it was possible to vary several of the properties of the goethite. At low temperature, the crystals were smaller and each crystal comprised many domains. Some of the properties varied continuously with an increase in temperature and those seemed to depend on the surface area. However other properties seemed to vary discontinuously with a sharp change near 50° C. These included: the *a*-dimension of the unit cell, the domain size in the *a* and *b* directions, and the strength of the H-bond between OH and O groups. Structural defects in the crystal were suggested as one possible explanation of these effects. If the crystals formed at low temperature were heated, the serrated ends of the crystal disappeared and the surface areas decreased. Further, the surface areas calculated from the crystal dimensions were similar to those calculated from adsorption — whereas before heating they had been much smaller. This suggested that micropores had disappeared. It seems likely that samples of goethite that have a low crystallinity may be more prone to continue to react with adsorbates but this does not appear to have been tested as yet.

There have been few detailed measurements of the long-term rate of reaction of adsorbates with goethite. One exception is the observation discussed above that removal of carbonate from goethite may be slow. In addition, both Madrid and Posner (1979) and Anderson *et al.* (1985) reported that a slow reaction occurred between goethite and phosphate and that it continued for several days. A more marked continuing reaction has been reported between metals and goethite (Gerth and Bruemmer 1981, 1983). It was especially marked for nickel and faster for zinc than for cadmium. Even so, there are many reports and much theory treating the reaction as if equilibrium had been reached after shorter periods. These will be considered first, and the question of rate of reaction deferred.

The reaction of cations with oxides

There are several published studies of the reaction of cations with oxides. Most give results like those in Fig. A2.1. Characteristically there is little reaction at a low pH and then, with

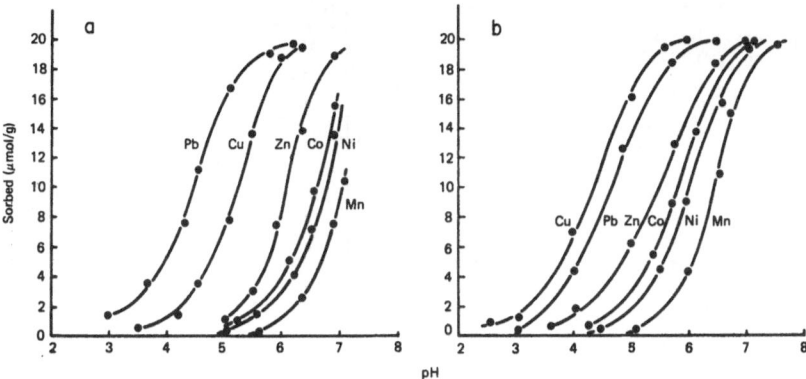

Fig. A2.1   Sorption of a range of materials on a) hematite and b) geothite when they were added at a rate of $20\mu mol/g$ of oxide. The values for the dissociation of metals to give the monovalent $MeOH^+$ ions are Pb,7.71; cu, 8; Zn 8.96; Co, 9.65; Ni, 9.86; and Mn, 10.59 (from McKenzie 1980).

increasing pH, a rapid transformation to almost complete adsorption. The pH at which reaction begins has been called the adsorption edge. It differs for different metals as shown in Fig. A2.1. This figure also shows that the sequence in which the metals adsorb may differ between adsorbates. Thus lead was adsorbed before copper on hematite but copper was adsorbed first on goethite. Most oxides with a high point of zero charge behave more-or-less as illustrated in Fig. A2.1. Because the point of zero charge is high, these oxides are positively charged at medium pH values. However, if the point of zero charge is low, as it is for manganese oxides, the charge is negative at medium pH values. There is then a much stronger tendency for reaction to occur at lower pH values (McKenzie 1980).

For the metals indicated in Fig. A2.1, the position of the adsorption edge increases as the pK_1 of the corresponding metal increases. However the adsorption edge occurs two to three pH units below the pK. This is partly a function of the experimental conditions because the position of the adsorption edge can be made to vary by varying the relative concentrations of adsorbate and adsorbent. The reasons for the relationship with the pK_1 will be considered in more detail in the chapters devoted to models. However, it is convenient to point out here, that one interpretation is that the species $MeOH^+$ reacts with the surface and that reaction occurs once this species has reached sufficient concentration. A difficulty in testing explanations of such observations is that the behaviour of many of the metals is so similar. Explanations can be tested more rigorously when there is diverse behaviour. It is therefore of interest to compare the behaviour of mercury. Mercury differs in that both the pK_1 and the pK_2 are low (Chapter A1). The species $HgOH^+$ therefore only occurs at low pH (Fig. A1.3).

The adsorption of mercury(II) onto finely-divided silica was studied by MacNaughton and James (1974). They found that, in the absence of chloride, adsorption began at about pH 2, reached a maximum at about pH 5 and then decreased (Fig. A2.2). The observation that adsorption decreases above pH 5 is important. If reaction were indeed due to the $HgOH^+$ species in solution this is the behaviour to be expected. The peak in adsorption

9

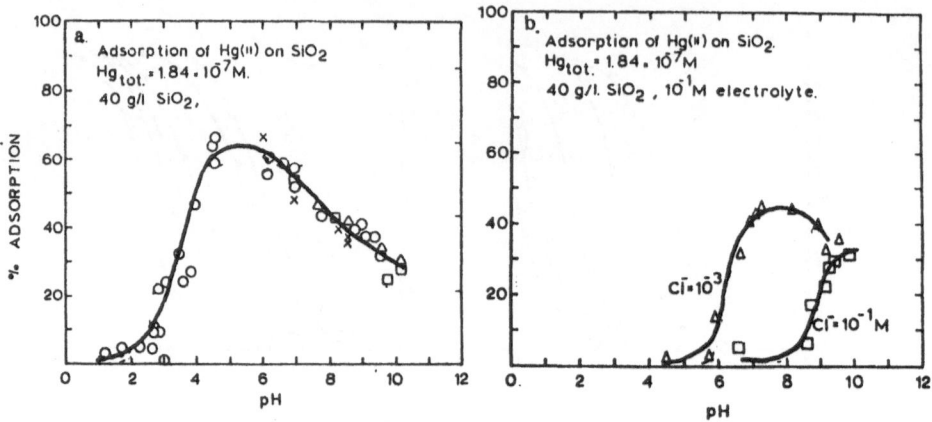

Fig. A2.2   The sorption of mercury by silica. In part a), open circles indicate sorption from sodium perchlorate and the other symbols sorption from magnesium nitrate solutions. In part b) sorption is from mixtures of sodium chloride and sodium perchlorate solutions (redrawn from MacNaughton and James 1974).

occurs after the peak in $HgOH^+$ abundance in solution because the decreased concentration of $HgOH^+$ is, at first, offset by the smaller (more negative) charge on the surface — thus favouring adsorption of positive ions. In the presence of chloride, there was little adsorption at low pH. This is consistent with other reports that, at low pH, chloride decreases adsorption of Mercury(II) on oxides (Forbes *et al.* 1974; Kinniburgh and Jackson 1978). However in the presence of chloride there was appreciable adsorption at higher pH. For 0.001 M chloride, the maximum retention was near pH 7 and for 0.1M chloride adsorption was apparently still increasing near pH 10 (Fig. A2.2). These pH values correspond to the peaks of the $HgOHC_1$ species (Fig. A1.3). Furthermore MacNaughton and James (1974) observed that adsorption of mercury(II) from nitrate solutions changed the surface potential (as indicated by electrophoretic mobility) towards more-positive values whereas adsorption from chloride solutions had little effect. These results are consistent with reaction with the positive $HgOH^+$ ion in the absence of chloride and with the neutral HgOHCl species in the presence of chloride. They suggest that the sheath of water molecules around the ion is weakened when one of the molecules loses a proton. The resulting species is then more prone to take part in ligand exchange reactions with oxides.

The reaction of anions with oxides

There are few studies in which the adsorption of a range of anions has been compared. The outstanding example is the work of Hingston *et al.* (1972). They showed that the effects of pH on adsorption were diverse. For some anions, for example fluoride and silicate, adsorption increased at first with increasing pH, and then decreased (Fig. A2.3). For others, for example phosphate and selenate, there was a general trend for adsorption to decrease as the pH increased but there were bends in the curves near some of the pK values of the relevant acid (Fig. A2.3). This behaviour was observed for the following adsorbates: fluoride, molybdate, tripolyphosphate, pyrophosphate, orthophosphate, selenite, silicate and arsenate.

10

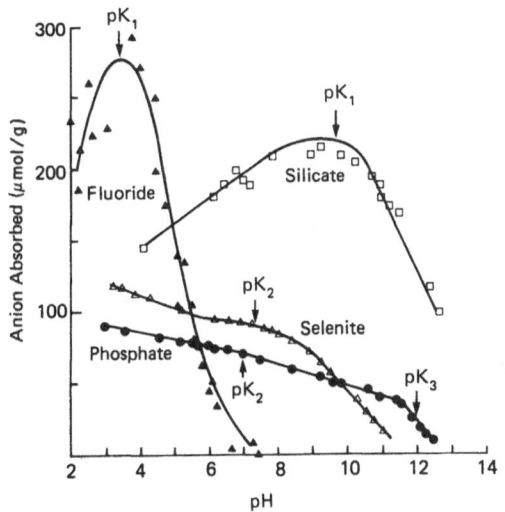

Fig. A2.3   Sorption of a range of anions on geothite. Two samples of geothite were used and the levels of addition differed for the differing anions (redrawn from Hingston *et al.* 1972).

Another characteristic of anion adsorption is that a well-defined maxmimum adsorption can often be specified (see Fig. A2.4). This maximum may occur at different pH values for each anion. (It is more difficult to specify a maximum adsorption for metals because the solubility of $Me(OH)_2$ is often low. As a result, at high concentrations of the metal in solution, precipitates may occur thus making it difficult to measure maximum adsorption.) Hingston (1981) has tabulated the apparent area occupied by one molecule of adsorbed anions on goethite and gibbsite. For example phosphate may occupy an area of $63\mathring{A}^2$ on goethite. This corresponds to a maximum adsorption of 2.6 μmol per m^2. Similar areas were occupied by arsenate, pyrophosphate and silicate. However sulfate, selenate and tripolyphosphate appear to require an area half as large again, whereas two molybdate molecules can fit into the area required for one phosphate molecule.

It is also characteristic of measurements of anion adsorption that results are often reported as graphs of amounts adsorbed versus concentration (Fig. A2.4). This probably stems from the interest of soil scientists in specifying the relation between adsorption and concentration. This relation indicates the buffering capacity of the soil for the nutrient in question and is important in determining both the rate of movement towards plant roots and the movement in leachate. Much effort has been expended in attempts to describe this relation and this will be considered later. Results for cation adsorption could, of course, also be presented in this manner but this is unusual. Indeed there are few studies in which a range of concentrations was used — for many of the reported experiments the only condition varied was the pH.

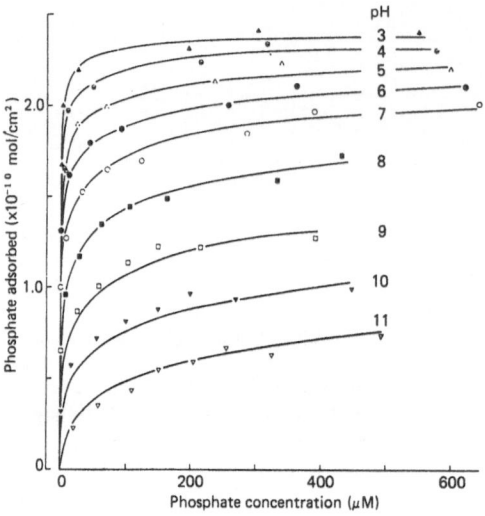

Fig. A2.4 Adsorption of phosphate by geothite at a range of pH values. The lines were fitted using the four-plane model described in Chapter A5 (from Bowden et al. 1980b).

Changes in charge resulting from adsorption

Whatever the mechanism of the reaction of ions with surfaces, the total charge in the system must be maintained. Consider the adsorption of zinc ions at a pH of say 7. At this pH, almost all of the zinc ions in solution will be divalent and so their average charge is close to 2. When adsorption of zinc ions occurs, two extreme possibilities could be envisaged. One is that all of the charge could be conveyed to the surface. The surface would then become more positive and the extra charge would be balanced by electrolyte anions closely approaching the surface (or by the repulsion of electrolyte cations). There would be no change in pH of the solution. The other extreme possibility is that the surface charge would not change as a result of adsorption. There would then be no tendency for extra electrolyte anions to approach the surface. Instead protons would be released in order to maintain the charge and, as a result, the pH of the solution would tend to fall. We are not concerned here with the source of these protons — merely that they must be produced. In practice neither of the extreme positions is reached. The charge on the adsorbing ions is balanced partly by release of protons and partly by changes in the concentration of electrolyte ions near the surface. The process could therefore be characterised either by specifying the release of protons per ion adsorbed or by specifying the mean change in surface charge per ion adsorbed. For cation adsorption it is more common to specify the protons released. Some representative values are given by Schindler (1981). He shows that values are usually between 1 and 2. Hence the charge conveyed to the surface is usually less than +1.

Much the same argument, but with opposite signs, can be made for adsorption of anions. In reporting such results, however, it is more common to specify the charge conveyed to the surface rather than the hydroxyls released. Some typical results are those of Hingston et al. (1974). They found, for example, values of about 0.6 for adsorption of phosphate on goethite and about 0.2 for adsorption of fluoride on goethite.

There is a tendency to regard such numbers as characteristic of the system of adsorbant and adsorbate. This is wrong. The values are affected by the pH, by the ionic strength of the background electrolyte, and by the amount of adsorption that has already taken place. Most data on these aspects is for anion adsorption. There is a tendency for the charge per ion adsorbed to be high at lower pH and to decrease in the middle range of pH. This is shown for the adsorption of phosphate on goethite in Fig. A2.5. This trend is also discernable for selenite adsorption on goethite (Hingston *et al.* 1972) and for silicate

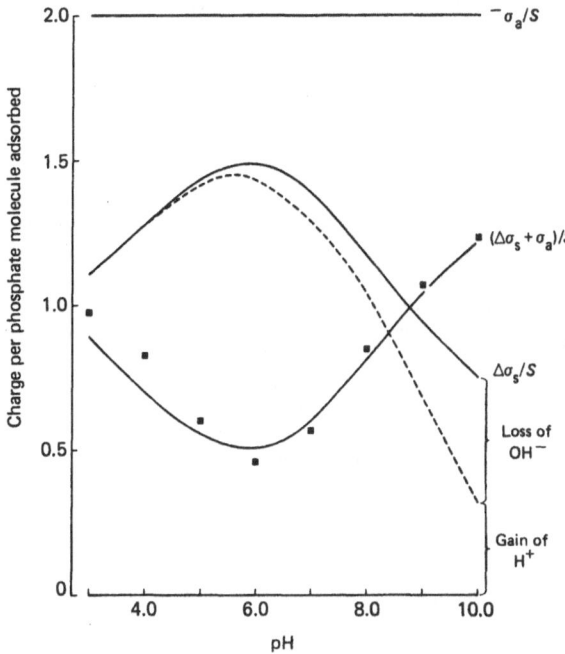

Fig. A2.5 The effect of pH on the charge conveyed to geothite by the adsorption of phosphate. The lines are derived from the four-plane model describedin Chapter A5. They indicate that the charge modelled is that due to adsorption of divalent HPO_4^{2-} ions less that due to displacement of OH^- ions and gain of H^+ ions (from Bowden *et al.* 1980b).

adsorption on goethite (Bowden *et al.* 1980a). However the trend is only marked in the presence of dilute electrolyte. A high concentration of background electrolyte gives a larger value and tends to flatten out the effects of pH.

Because the adsorption of an ion changes the charge on the surface, adsorption of each additional ion is onto a surface that is electrically different from that for the preceding ion. The charge on the ion may therefore be balanced differently. For example, for phosphate in the middle pH range, the charge per ion adsorbed is large at first, decreases to a minimum, and then increases again (Bolan and Barrow 1984).

Summary

A range of oxides occurs in soils but these oxides are far from pure. Because they are difficult to extract from soil, synthetic pure oxides have been studied as models. The reactions of cations and anions with such oxides have usually been measured over a specified time and the results treated as equilibrium adsorption. The behaviour of many cations is similar in that adsorption increases with increasing pH. Mercury is an exception; its behaviour is consistent with the adsorption of $MgOH^+$ and MgOHCl species. The behaviour of anions is more complex. The reaction with both anions and cations changes the charge on the surface and the pH of the solution but simple stoichiometry seldom occurs. In the next chapter, possible explanations of this behaviour are considered.

References

Atkinson, R.J., Posner, A.M., and Quirk, J.P. 1968. Crystal nucleation in Fe(III) solutions and hydroxide gels. Journal of Inorganic and Nuclear Chemistry 30, 2371-2381.

Anderson, M.A., Tejedor-Tejedor, I. and Stanforth, R.R. 1985. Influence of aggregation on the uptake kinetics of phosphate by geothite. Environmental Science and Technology 19, 632-637.

Benjamin, M.M. and Leckie, J.O. 1981. Multiple site adsorption of Cd, Cu, Zn and Pb on amorphous iron oxyhydroxide. Journal of Colloid and Interface Science 79, 209-221.

Bolan, N.S. and Barrow, N.J. 1984. Modelling the effect of adsorption of phosphate and other anions on the surface charge of variable charge oxides. Journal of Soil Science 35, 273-281.

Bowden, J.W., Posner, A.M. and Quirk, J.P. 1980a. Adsorption and charging phenomena in variable-charge soils. In Theng, B.K.G. (Editor), Soils with Variable Charge. New Zealand Society of Soil Science. Lower Hutt. New Zealand.

Bowden, J.W., Nagarajah, S., Barrow, N.J., Posner, A.M. and Quirk, J.P. 1980b. Describing the adsorption of phosphate, citrate and selenite on a variable-charge mineral surface. Australian Journal of Soil Research 18, 49-60.

Evans, T.D. 1976. Interfacial electrochemistry of goethite (α-FeOOH) PhD Thesis, University of Newcastle upon Tyne.

Evans, T.D., Leal, J.R. and Arnold, P.W. 1979. The interfacial electrochemistry of goethite (αFeOOH) especially the effect of CO_2 contamination. Journal of Electroanalytical Chemistry 105, 161-167.

Forbes, E.A., Posner, A.M. and Quirk, J.P. 1974. The specific adsorption of inorganic Hg(II) species and Co(III) complex ions on goethite. Journal of Colloid and Interface Science 49, 403-409.

Fordham, A.W., Merry, R.H. and Norrish, K. 1984. Occurrence of microcystralline goethite in an unusual fibrous form. Geoderma 34, 135-148.

Fordham, A.W. and Norrish, K. 1979a. Electron microprobe and electron microscope studies of soil clay particles. Australian Journal of Soil Research 17, 283-306.

Fordham, A.W. and Norrish, K. 1979b. Arsenate-73 uptake by components of several acid soils. Australian Journal of Soil Research 17, 307-316.

Gerth, J. and Bruemmer, G. 1981. Einfluss von Temperatur und Reaktionzeit auf die Adsorption von Nickel, Zink, und Cadmium durch Goethite. Mitteilungen der Deutschen Bodenkundlichen Gesselschaft 32, 229-238.

Gerth, J. and Bruemmer, G. 1983. Adsorption und Festlegung von Nickel, Zink und Cadmium durch Goethite (αFeOOH). Fresenius Zeitschrift fur Analytical Chemistry 316, 616-620.

Hingston, F.J. 1970. Specific adsorption of anions on goethite and gibbsite. PhD Thesis, University of Western Australia.

Hingston, F.J. 1981. A review of anion adsorption. In Anderson, M.A. and Rubin A.J. (Editors). Adsorption of Inorganics at Solid-liquid Interfaces. Ann Arbor Science Publ. Ann Arbor. Michigan.

Hingston, F.J., Posner, A.M. and Quirk, J.D. 1972. Anion adsorption by goethite and gibbsite. 1. The role of the proton in determining adsorption envelopes. Journal of Soil Science 23, 177-192.

Hingston, F.J., Posner, A.M. and Quirk, J.P. 1974. Anion adsorption by goethite and gibbsite. 2. Desorption of anions from hydrous oxide surfaces. Journal of Soil Science 25, 16-26.

James, R.D., Stiglich, P.J. and Healy, T.W. 1980. The TiO_2/aqueous electroyte system — applications of colloid models and model colloids. In American Chemical Society symposium series, Adsorption from Aqueous solution.

Kinniburgh, D.G. and Jackson, M.L. 1978. Adsorption of mercury (II) by hydrous oxide gel. Soil Science Society of America Journal 42, 45-47.

Kinniburgh, D.G. and Jackson, M.L. 1981. Cation adsorption by hydrous metal oxides and clay. In Anderson, M.A. and Rubin, A.J. (Editors). Adsorption of Inorganics at Solid-liquid Interfaces. Ann Arbor Science Publ.

Madrid, L. and de Arambarri, P. 1978. Adsorption isotherms and hysteresis of proton adsorption by goethite. Geoderma 21, 199-208.

Madrid, L., Posner, A.M. 1979. Desorption of phosphate from goethite. Journal of Soil Science 30, 697-707.

McKenzie, R.M. 1980. The adsorption of lead and other heavy metals in oxides of manganese and iron. Australian Journal of Soil Research 18, 61-73.

MacNaughton, M.G. and James, R.O. 1974. Adsorption of aqueous mercury(II) complexes at the oxide/water interface. Journal of Colloid and Interface Science 47, 431-440.

Norrish, K. and Rosser, H. 1983. Mineral phosphate. In CSIRO Div. of Soils (Editors), Soils: an Australian Viewpoint. CSIRO Melbourne/Academic Press.

Schindler, W. 1981. Surface complexes at oxide-water interfaces. In Anderson, M.A. and Rubin, A.J. (Editors). Adsorption of Inorganics at Solid Liquid Interfaces. Ann Arbor Science Publ. Ann Arbor. Michigan.

Schwertmann, U. 1985. The affect of pedogenic environments on iron oxide minerals. Advances in Soil Science 1, 171-200.

Schwertmann, U., Cambier, P., and Murad, E. 1985. Properties of goethite of varying crystallinity. Clays and Clay Minerals 33, 369-378.

Smith, K.L. and Eggleton, R.A. 1983. Botryoidal goethite: a transmission electron microscope study. Clays and Clay Minerals 31, 392-396.

Taylor, R.M., McKenzie, R.M., Fordham, A.W. and Gillman, G.P. 1983. Oxide minerals. In CSIRO Div. of Soils (Editors), Soils: an Australian Viewpoint. CSIRO Melbourne/Academic Press.

Taylor, R.M. and McKenzie, R.M. 1966. The association of trace elements with manganese minerals in Australian soils. Australian Journal of Soil Research 4, 29-39.

Yoon, R.H., Salman, T. and Donnay, G. 1979. Predicting points of zero charge of oxides and hydroxides. Journal of Colloid and Interface Science 70, 484-493.

Chapter A3

Describing and explaining the adsorption behaviour of oxides

The differences between describing and explaining

The title of this chapter includes both "describing" and "explaining". The distinction is not always clear, yet it can be important. It is possible to describe a set of observations in a completely non-explanatory manner. An example is the use of a quadratic function to describe a simple curve. In most cases such descriptions are non-mechanistic. They can be of value for summarizing a large set of observations by a few numbers. They are also of value for interpolation but of limited value for extrapolation. However, if we want to understand the processes involved, and thus explain the observations, we need a different approach. If this understanding is formalized into sets of relations, and these relations are precisely expressed as equations, we may call it a mathematical model. Thus a mathematical model is no more than a precisely expressed hypothesis. The confusion arises when we wish to test that hypothesis. This involves testing whether it *describes* the observed behaviour. Thus in both cases the test is the same. What then are the characteristics of an explanatory or mechanistic model? One is that it contains elements of an underlying mechanism. However mechanisms seem to be like Russian dolls — there is always another one inside. What seems mechanistic to one may seem superficial and descriptive to another. Thus judging on this criteria is partly subjective. Another characteristic of a mechanistic description is that it is comprehensive. There are several ways of describing an elephant's trunk — but a comprehensive model is also capable of describing the rest of the elephant. The criteria of comprehensiveness is part of the normal scientific process of testing hypotheses/models; if a model is comprehensive, we should be able to make predictions and to test them.

Three main models have been used to describe the adsorption behaviour of oxides. They have been compared by Barrow & Bowden (1987). In the text that follows, I have drawn heavily on this work and, in many cases, have used the same equations.

The models range across the spectrum from fairly descriptive to fairly mechanistic and from single-purpose to comprehensive. All three simplify the distribution of molecular species near a reacting surface by allocating them to mean planes of adsorption. The word "mean" needs emphasis. When more than one plane is specified, it is not envisaged that all the protons line up in one plane and all the chloride ions in another. This would imply a far too regular surface to the oxide. Rather it is envisaged that, on the average, the protons tend to occur closer to the metal atoms of the oxide than the chloride ions and therefore "inside" them. There may well be inter-penetration of the distributions of the ions so that the mean planes are closer than ion dimensions. The allocation to a mean plane is a device by means of which the effects due to the charge on the surface and on the ions, can be described. Thus the mean plane of adsorption is assumed to also have a certain electric potential, and the ions in that plane are all assumed to experience that potential. The models differ in the number of planes that are specified and this provides a convenient labelling system.

The single plane model

The simplest model specifies only one plane of adsorption (Fig. A3.1). All ions that are specifically adsorbed are allocated to this plane and hence experience the same potential.

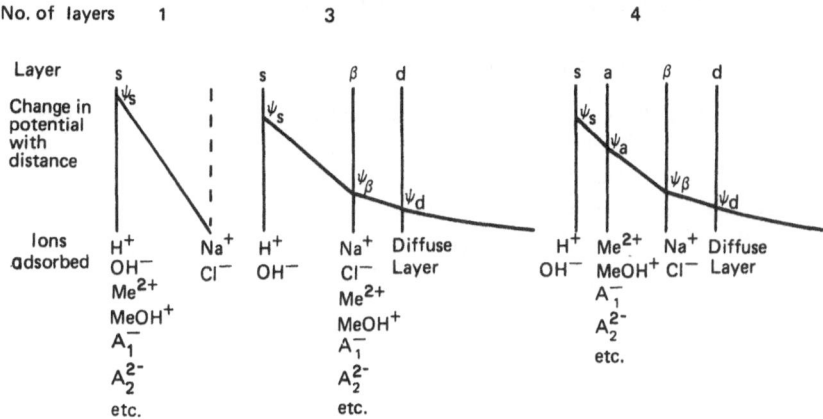

Fig. A3.1   Diagrammatic representation of the arrangement of the postulated mean planes of adsorption of the three models.

Thus protons, hydroxyl, sodium, chloride and phosphate ions are all allocated to this one plane. No explicit account is taken of the background electrolyte either by specific interactions with surface sites or in terms of the diffuse layer.

This model is also the oldest as it was used by Gilbert and Rideal (1944) to describe the charge on wool fibres. It was subsequently used by Atkinson et al. (1967) for oxide systems and has also been used more recently by Sigg and Stumm (1980) and by Goldberg and Sposito (1984). It corresponds to the "first model" of Morel et al. (1981). It has been called the "constant capacitance" model by Westall and Hohl (1980) and by Goldberg and Sposito (1984). This term apparently arose because the electric potential of the surface is assumed to be proportional to the charge and thus the capacitance is assumed to be constant. However virtually all models end up with a simple proportionality between potential and charge (Morel et al. 1981). Further, if the background electrolyte is changed, this is accommodated in the model by a change in the value for capacitance (Westall and Hohl 1980). This term therefore seems unsuitable.

In this model the surface sites are treated like ordinary chemical entities and the aim is to write equations to describe the reactions which are thought to take place at the surface. Thus the reactions with protons can be written as an ordinary dissociation reaction:

$$SOH_2^+ \rightleftharpoons SOH + H^+ \qquad\qquad\qquad A3.1$$

and

$$SOH \rightleftharpoons SO^- + H^+ \qquad\qquad\qquad A3.2$$

Equations for the reaction with absorbed ions may also be written. For example, for phosphate, Goldberg and Sposito (1984) wrote:

$$SOH + H_3PO_4 \rightleftharpoons SH_2PO_4 + H_2O \qquad\qquad\qquad A3.3$$

$$SOH + H_3PO_4 \rightleftarrows SHPO_4^- + H_2O + H^+ \qquad\qquad A3.4$$

$$SOH + H_3PO_4 \rightleftarrows SPO_4^{2-} + H_2O + 2H^+ \qquad\qquad A3.5$$

where groups prefixed by an S represent surface sites. Thus, in this case, three surface species are proposed. These species may be regarded as dissociating so that, at low pH the uncharged SH_2PO_4 dominates and, as the pH increases, the singly charged $SHPO_4^-$ and then the doubly charged SPO_4^{2-} are found.

Because the reaction is with a charged surface, the equilibrium expressions for these reactions are more complex than for an ordinary reaction in solution. For example, for equation A3.1

$$K_1 = \frac{[SOH][H^+]\exp(-F\psi/RT)}{[SOH_2^+]} \qquad\qquad A3.6$$

and for equation A3.5

$$K_3^1 = \frac{[SPO_4^{2-}][H^+]^2\exp(-2\,F\psi/RT)}{[SOH][H_3PO_4]} \qquad\qquad A3.7$$

while the 'K' terms indicate intrinsic conditional equilibrium constants (for these constants, superscripts are used here to indicate the number of layers of the model), the square brackets indicate concentrations, ψ the electric potential in the plane of adsorption, F the Faraday constant, R the gas constant and T the absolute temperature. Analogous equations could be written for the other assumed reactions. The exponential terms in these equations take into account the electric pontential in the plane of adsorption. They are included for those reactions for which an ion moves onto, or from, the charged surface. According to Goldberg and Sposito (1984) the exponential terms can be considered as solid-phase activity coefficients that correct for the charge of a surface species. Their effect is to modify the influence of the equilibrium constants. Thus equation A3.7 could be written

$$K_3^1\exp(2F\,\psi/RT) = \frac{[SPO_4^{2-}][H^+]^2}{[SOH][H_3PO_4]} \qquad\qquad A3.8$$

If the value of ψ is negative, the effect of this term is to decrease the value of the left hand side of the equation. Thus it decreases the amount of product formed. This is to be expected because protons would not escape readily from a negative surface. Hence, in this model, there are two opposing effects of pH. High pH favours reaction A3.5 on simple, mass-action grounds, but the negative charge on the surface at high pH would tend to decrease the importance of the reaction.

With such models there are two steps in finding how the model behaves. One is allocating values to the parameters K_1^1, K_2^1, etc., and the other is solving the system of equations. Suppose we have a set of provisional values for the parameters. Then we can treat the assumed complexes as in an ordinary chemical reaction. The only difference is that we have to assign a value to the potential ψ. This may be obtained by an iterative process. If we assume an initial value of ψ, the equations can be solved and the surface charge may be calculated by summing the charged species. The potential corresponding to this charge can be calculated from:

$$\psi = \sigma \, F/c \qquad\qquad\qquad A3.9$$

where σ is the surface charge in mol. cm^{-2} and c is the capacitance density in Farad m^{-2}. If the initial value of ψ had been correct then the value of ψ calculated from equation A3.9 would be the same as the initial value. Solving the equations therefore involves systematically varying the initial value of ψ until it agrees with the value calculated from equation A3.9. Allocating values to the parameters of the model may involve a mixture of techniques. For example Goldberg and Sposito (1984) allocated values to the parameters K_1 and K_2 from a consideration of values used in previously published reports. However the values for the parameters K_1^1, K_2^1 and K_3^1 can only be allocated on the basis of matching the existing data. A program such as SIMPLEX (Part B) can be used to systematically vary the values of these parameters until the ones that best fit the data are found.

In evaluating this model we may use three criteria. The first is to judge how effective the model is in describing adsorption. Under restricted conditions it can be quite effective. For example, for phosphate, it describes the effect of increasing pH on adsorption by summing the separate effects of pH on each of the three supposed products as indicated in equations A3.3-A3.5. Each of the three separate reactions, when considered individually gives an inappropriate curve for the effects of pH. However, when added together the result is a flexible curve that can reproduce the general effects of pH fairly well (Fig. A3.2). It also describes the effects of phosphate concentration on adsorption fairly well (Fig. A3.2, A3.3).

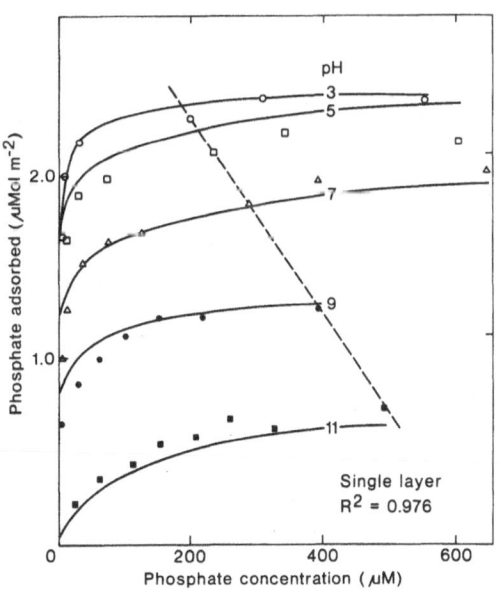

Fig. A3.2   Effects of pH and of phosphate concentration on phosphate adsorption as described by the single-plane model. The broken line represents one level of addition of phosphate. The fit in the pH direction at this level of phosphate is shown in Fig. A3.3. Date at pH 4, 6, 8, and 10 were omitted for clarity (from Barrow & Bowden 1987).

19

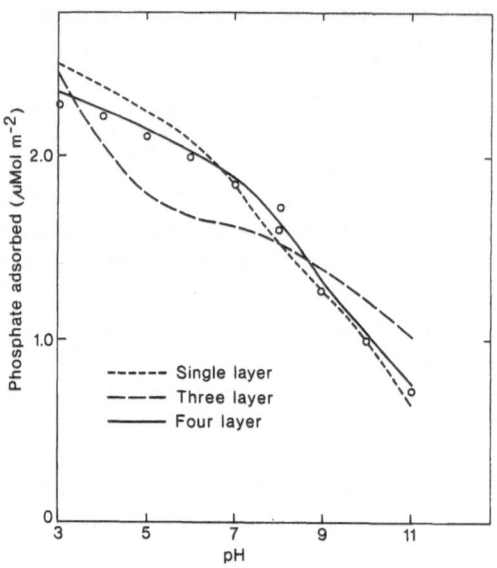

Fig. A3.3 Effect of pH on the adsorption of phosphate as described by the three models (from Barrow & Bowden 1987).

The second criteria is its comprehensiveness. The model scores poorly on this criteria. It is not comprehensive — rather it can be considered as a set of different models each of which would be used for different conditions. For example, the model has no facility to take into account the ionic strength of the background electrolyte. This can only be done by arbitrary changes to the value of the capacitance term (c of equation A3.9). As there are no rules provided for these changes, the result is a separate model for each ionic strength. The model also requires a specification of the concentration of sites. These sites may then react with protons or with phosphate ions. As a result, the maximum surface concentration of charge, either positive or negative, and the maximum surface concentration of phosphate are specified as being the same. This is not consistent with observations. As a result, the model can only work for phosphate at low ionic strengths because only at low ionic strength is the maximum surface charge small. This problem becomes even more acute when dealing with a small ion such as fluoride that can adsorb to far higher densities than phosphate or indeed with other ions which adsorb to differing maximum concentrations (Chapter A2). It is not possible to have a common maximum value for charge and for adsorption of all anions.

The third criterion for evaluating the model is realism. This is at least partly subjective. However, it seems to me to be unrealistic to ignore ion size and to postulate that such diverse ions as H^+, HPO_4^{2-}, F^-, and Cl^- can all be allocated to the same plane of adsorption. It is because considerations of ion size are ignored that a common maximum adsorption for phosphate and for protons is postulated as noted above. It is also questionable to postulate that phosphate forms only monodentate links to the surface. There is appreciable evidence that phosphate forms binuclear bonds with the surface. Goldberg and Sposito (1985) have reviewed this evidence and have concluded that studies based on isotopic exchange and on infrared spectroscopy are ambiguous. Nevertheless they also show from crystallographic calculations that it has not proved possible to fit enough phosphate molecules onto a goethite surface to comply with a monodentate link.

The maximum number is consistent with a binuclear complex. It therefore seems unsafe to construct a model that requires monodentate binding and to then restrict the amount of adsorption produced by the model by choosing a small number for the maximum number of sites.

The three-plane model

This model differs from the preceding one in the number of planes that are specified (Fig. A3.1). It also differs in that specific account is taken of electrolyte ions both in forming complexes with the surface and in forming a diffuse layer. It has been called the "triple-layer" model by Westall and Hohl (1980) and is the "third model" of Morel *et al.* (1981). The model has been applied to several sets of results and has been described in several review articles. In the present instance, we will consider its application to adsorption of metal ions and to anions as described by Davis and Leckie (1978, 1980).

In this model, sites are defined as in the single-plane model and hence equations A3.1 and A3.2, and the equilibrium equations derived from them, are common to both. In addition, it is assumed that surface complexes are formed with electrolyte ions. For example:

$$SO^- + Na^+ \rightleftarrows SO^-\!\!-\!Na^+ \qquad\qquad A3.10$$

Note that the charges are not pictured as being neutralized and the complex is envisaged as retaining a negative charge in the first plane and a positive charge in the second plane (Fig.1). Because the electrolyte ions are assigned to the second plane, the exponential term in the equilibrium equation contains a term for the electric potential in this plane (ψ_β). Thus for sodium:

$$K_{Na} = \frac{[SO^-\!\!-\!Na^+]}{[SO^-][Na^+]\exp(-F\psi_\beta/RT)} \qquad\qquad A3.11$$

In the derivation of these equations the exponential term was introduced using the Bolzmann distribution. Sposito (1983) has argued that this is both unnecessary and undesirable and that the exponential terms should be considered as solid-phase activity coefficients.

Analogous equations would be written for the electrolyte anion. It is an important characteristic of the implementation of this model by Davis and Leckie (1978, 1980) that the intrinsic equilibrium constants for electrolyte ions are large and hence that much of the charge on the first plane is balanced by these complexes rather than by charge in the diffuse layer.

Reaction with specifically adsorbed ions is treated in an analogous way to that for the electrolyte ions. All applications have involved divalent ions and it has been assumed that both divalent and monovalent surface species are formed. For a divalent anion A^{2-} the equations would be written:

$$SOH + H^+ + A^{2-} \rightleftarrows SOH_2^+\!\!-\!A^{2-} \qquad\qquad A3.12$$

21

and

$$SOH + 2H^+ + A^{2-} \rightleftharpoons SOH_2^+ {-} HA^-$$ A3.13

For these equations, two species of ions are moved from the solution to the charged surface and so the equilibrium expressions are:

$$K_1^3 = \frac{[SOH_2^+ {-} A^{2-}]}{[SOH\ H^+]\,[A^{2-}]\,\exp\,((2\psi_\beta - \psi_0)\,F/RT)}$$ A3.14

and

$$K_2^3 = \frac{[SOH_2^+ {-} HA^{2-}]}{[SOH]\,[H^+]^2\,[A^{2-}]\,\exp\,((\psi_\beta - \psi_0)\,F/RT)}$$ A3.15

Thus, for this model the effects of pH on adsorption are attributed to the different but additive effects of pH on the proposed two separate reactions. That is, this model also adds the curves arising from the proposed separate reactions. The individual curves do not give the correct effect of pH but when added together they may approximate to the correct curve.

The model differs further from the single-plane model in that the charge in the diffuse layer is calculated from the Gouy-Chapman theory. A complete list of the equations used is given in the original papers and in Morel *et al* (1981).

As for the single-plane model, it is assumed that the surface species are mutually exclusive and hence a mass balance can be used to sum the sites. Solution of the equations is more complicated because there are two electric potentials to be estimated. A clear exposition of the iterative methods that have been used is given by Morel *et al*. (1981).

A characteristic of both the single-plane and the three-plane models is that the effects of pH on adsorption are allocated partly to mass action effects of H^+ concentration and partly to electrostatic effects via the potential of the surface. Thus the mass action effects of H^+ would tend to favour reaction A3.13 over A3.12 and so give rise to different effects of pH on the two proposed reactions. However these effects are modified by the electrostatic potential. For the three-plane models the magnitude of the effect of potential depends on the value allocated to the capacitance and hence to the difference between ψ_β and ψ_0.

The three-plane model may be judged on the same criteria as the single plane model. The first of these is its ability to describe adsorption. The three-plane model only seems to work well when the effects to be described are simple. Thus Davis and Leckie (1980) found that the model described the effects of pH on adsorption of sulfate and selenate. For both of these ions, adsorption decreases in a simple way as pH increases. Similarly the model described the effects of pH on adsorption of metals (Davis and Leckie 1978). In this case adsorption increases in a simple way with increasing pH. These simple curves are modelled by adding the separate effects of the two assumed reactions. The model fails to describe adsorption under two separate conditions. One is when the adsorbing ion is monovalent — for example, fluoride. In this case it is not possible to postulate products analogous to those in A3.12 and A3.13 and the model describes adsorption poorly (Barrow and Bowden 1987). The other case is when the effects of pH are more complex, as, for example, for phosphate. Figures A3.3 and A3.4 show that the model describes phosphate adsorption poorly.

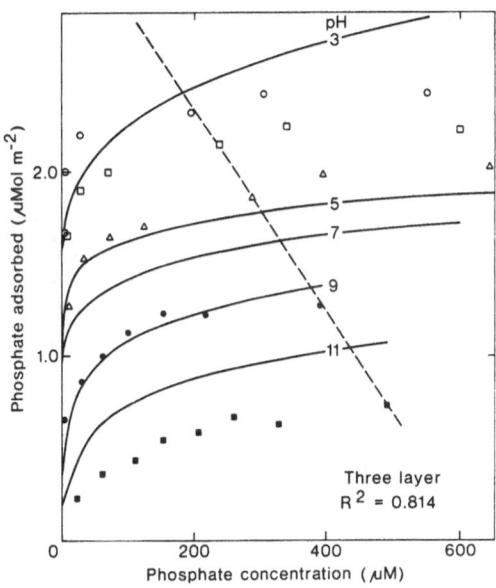

Fig. A3.4　Effects of pH and of phosphate concentrate on phosphate adsorption as described on the three-plane model. Compare Fig. A3.2 (from Barrow & Bowden 1987).

The three plane model scores better than the single-plane model in terms of comprehensiveness. It is designed to be able to accommodate the effects of changing the ionic strength and to be able to reproduce effects on electrophoretic mobility. However, as it can only closely describe the adsorption of a few ions, it cannot be considered to be a comprehensive model.

On the scoreboard for realism there are three counts against the model. One is the problem posed by the fact that anions differ in the closeness of packing they can reach and thus on their maximum adsorption density (Chapter A2). This problem is not troublesome for adsorption of metals because precipitation of the hydroxide usually occurs well before the maximum adsorption is reached. The models therefore are not required to reproduce sorption close to the maximum. The problem with anions was recognised by Davis and Leckic (1980). They attempted to overcome it by arguing that an anion may "cover" more sites than it is bonded to. For example a selenate ion may "cover" three sites even though it is postulated to be bound to only one. These "covered" sites are then allocated to the selenate complex in the mass-balance equation. This means they are removed from other possible reactions including protonation-deprotonation reactions. This seems to be a serious weakness of the model. Furthermore it is a basic weakness as it arises from treating "sites" as if they were indepedent chemical entities.

A second count against the model is that it allocates very different ions to the same plane of potential. This means that F^-, HPO_4^{2-} and Cl^- ions are considered to all experience the same change in electric potential with change in pH. This seems unlikely.

A third count against the model is that it is necessary to specify the reactions and reaction products that are formed.

For example, for a divalent anion, the following set of products seems feasible: SOH_2^+ -HA^-, SOH-HA^-; SOH_2^+-A^{2-}; SOH-A^{2-}; $(SOH)_2^{2+}$—A^{2-}; and $(SOH, SOH_2)^+$—A^{2-}. Others have been omitted on electrostatic grounds. The modeller then has the choice of either selecting a limited set from these possibilities or of including all possibilities. If all the possibilities were included there would be so many parameters that the exercise could be regarded as mere curve-fitting.

The four-plane model

An important assumption of the model is that certain ions react with the surface by ligand exchange. Thus, fluoride and anions of oxyacids are thought to replace surface hydroxyl and water molecules. It is therefore logical to expect such ions to take up a position on the surface different to that of ions that form outer-sphere complexes. They are therefore allocated to a separate plane. Because of mathematical complexity, it is difficult to introduce a further plane using the surface speciation approach of the three layer model (Davies and Leckie 1978). For this reason, and because of the problems with large ions as described above, surface speciation was avoided. Instead it is assumed that the surface activity of an ion is proportional to the ratio of occupied sites to vacant sites.

It is further assumed that, on the average, a neutral surface site can be written as:

$$S\!\!<^{\displaystyle OH}_{\displaystyle OH_2}$$

This specification arises from a consideration of the stoichiometry of the metal atoms near the surface of a metal oxide. It has been common in the soil science literature for some time (eg (White 1980)) and is now appearing in the surface science literature (Pulver $et\ al.$ 1984).

Protonation and deprotonation can be specified as before:

$$S\!\!<^{\displaystyle OH_2}_{\displaystyle OH_2}\Big]^{+} \;\rightleftarrows\; S\!\!<^{\displaystyle OH}_{\displaystyle OH_2}\Big]^{\circ} +H^{+} \tag{A3.16}$$

$$S\!\!<^{\displaystyle OH}_{\displaystyle OH_2}\Big]^{\circ} \;\rightleftarrows\; S\!\!<^{\displaystyle OH}_{\displaystyle OH}\Big]^{-} +H^{+} \tag{A3.17}$$

and thus K_1 and K_1 can be defined analogously. However, in explaining a characteristic of this model, it is instructive to write the equations as:

$$S\!\!<^{\displaystyle OH}_{\displaystyle OH_2}\Big]^{\circ} +H^{+} \overset{K_H}{\rightleftarrows} S\!\!<^{\displaystyle OH_2}_{\displaystyle OH_2}\Big]^{+} \tag{A3.18}$$

$$S\!\!<^{\displaystyle OH}_{\displaystyle OH_2}\Big]^{\circ} +OH^{-} \overset{K_{OH}}{\rightleftarrows} S\!\!<^{\displaystyle OH}_{\displaystyle OH}\Big]^{-} \tag{A3.19}$$

This was the form used by Bowden $et\ al.$ (1977). When these are the only reactions considered, the process can be regarded as a competition for surface sites and the expression derived by Bowden $et\ al.$ (1977) has the form of a competitive Langmuir equation:

$$\sigma_s = \frac{N_S\ [K_H\ a_H\ exp\ (-F\psi_S/RT) - K_{OH}\ a_{OH}\ exp\ (F\psi_S/RT)}{1 + K_H\ a_H\ exp(-F\psi_S/RT) + K_{OH}\ a_{OH}\ exp(F\psi_S/RT)} \qquad A3.20$$

where σ_s is the charge due to reactions A3.18 and A3.19, a indicates activity and N_S is the site density.

In the derivation (Bowden $et\ al.$ 1977) of equation A3.20, the exponential terms were introduced from a consideration of the activity of an ion at the charged surface.

Consider now the reaction of the surface site with a monovalent anion A_1^-. By analogy with equation A3.19 we might write:

$$S \underset{\diagdown OH_2}{\overset{\diagup OH}{\bigg]}}^{\circ} + A_1^- \rightleftarrows S \underset{\diagdown A_1}{\overset{\diagup OH}{\bigg]}}^- + H_2O \qquad A3.21$$

Alternatively the A_1^- ion could be regarded as replacing the surface hydroxyl and thus producing no change in charge. Because an anion can replace either a water molecule or a hydroxyl ion, it can react with a surface site no matter what its surface charge. That is, in this context, all sites are regarded as vacant. Because all sites are vacant, the charges on the reacting sites are not specified. Instead the ions are thought of as reacting with sites of average charge and potential for the given conditions of pH and ionic strength. Further, separate equations are not used to specify whether the charge on the adsorbing ion is balanced as in equation A3.21 or by displacing a surface hydroxyl. Instead this is also determined by the electric potential of the surface. This specification differs in concept from that of the other models. It has the advantage that it permits the introduction of a term for maximum adsorption the value of which can differ for different adsorbed ions. Thus for reaction with the anion A of valency z, the equilibrium equation is:

$$K_z^4 = \frac{\Gamma_A^{z-}}{[N_T - \Gamma_A z-]\ [A^{z-}]\ exp\ (zF\psi_a/RT)} \qquad A3.22$$

where Γ_A^{z-} is the surface excess concentration of A^{z-} and N_T is the maximum value of $\Gamma_A z-$. Comparison with equations A3.6, A3.7, A3.14 and A3.15 shows that there are analogies between the constants of this model and those of the other models but that they differ because of the way the reaction is specified.

For a polyvalent adsorbate, the model permits reaction with all of the species of ion present in solution. However, in practice, it is found that one species usually dominates. Thus, for this model, the subscript to the "K" term indicates the valency of the adsorbing ion. Hence the model usually involves a choice between known species in solution rather than a choice amongst proposed surface species.

The ratio N_S/N_T gives the mean number of sites occupied by the ion A - . Because the possible packing of ions on a surface is not known with certainty, this ratio is not constrained to be an integer. Equation A3.22 was originally derived by setting the ratio of occupied to vacant sites equal to the surface activity of the ion A^-. It has also been derived from a kinetic approach in which both the velocity and the end point of the reaction are shown to depend on the electric potential of the surface (Barrow *et al.* 1981 and Chapter B3). Chapter B3 shows how this kinetic approach may be used to derive equations for the rate of adsorption and for equilibrium adsorption when either the starting conditions or the final concentration is specified.

Equations analogous to equation A3.22 are used for the ions in the β plane and the charge in the diffuse layer is calculated from the Gouy-Chapman theory. The model is completed by equations interrelating the charge and the potential via the capacitance and by a charge-balance equation. A list of the equations is given in (Table B2.1). The equations of the model are solved using an iterative method.

An important characteristic of the model is that the effects of pH are allocated to two separate effects on the one reaction. One of these effects is on the electrostatic potential of the surface; the other is on the ions present in solution. The effects of potential are mediated through the exponential term of equation A3.22. The effect is that adsorption of anions is favoured when the potential is positive and discouraged when it is negative. That is, this term has the effect of decreasing anion adsorption as the pH increases. On the other hand it increases cation adsorption as the pH increases. However, adsorption is determined by the product of: the "K" term, the concentration term, and the electrostatic term. If the other terms are large enough, appreciable adsorption of anions can occur on negative surfaces, and appreciable adsorption of cations on positive surfaces. Because the concentration term refers to the concentration of an *ion*, its magnitude will depend on the degree of dissociation. Consider an anion for which the pK of the relevant acid is in the mid range. Then, as the pH is increased towards the pK, the concentration of the ion will increase. This effect will tend to increase adsorption and thus oppose the effects of the increase in pH on the electric potential of the surface. However, once the pK is exceeded, there is little scope for further increases in the concentration of the ion and so the effect of the electric potential dominates. There is therefore a change in slope of plots of adsorption versus pH — as is observed (Chapter 2). Consider now a cation. If we assume that the $MeOH^+$ ions are adsorbed, then, as the pH rises, the increasing concentration of these ions and the decreasing (or more negative) charge on the surface have a complementary effect. As a result, adsorption increases rapidly with increasing pH once a certain "threshold" value is reached — again, as is observed (Chapter 2). This model therefore has a simple and consistent way of explaining the effects of pH on both anion and cation adsorption. This is described in more detail in Chapter A5.

The four-plane model scores very well on the criterion of ability to describe observations. Thus it has been shown to be able to very closely describe: the adsorption of phosphate, selenite and citrate (Bowden *et al.* 1980); the effects of electrolyte concentration on adsorption of phosphate (Barrow *et al.* 1980); the adsorption of copper, lead and zinc (Barrow *et al.* 1981); and the adsorption of fluoride, sulfate and silicate (Barrow and Bowden 1987). In several of these studies, the model was tested not only against the effects of pH but also against the effects of varying the concentration. This is illustrated for phosphate in Figures A3.3 and A2.4. In addition, the four-plane model was able to closely describe the effects of adsorption on charge both for anions (Bowden *et al.*1980 and

Fig. A2.5) and for cations (Barrow *et al.*1981). Not only is it able to describe the observations closely, it is also able to describe them efficiently — that is, using few parameters. When required to describe charge in the absence of adsorption, both the three-plane and the four-plane model require the same number of parameters. This is because the physical structure of the models is similar and they differ only in the way the equations are written. When required to also describe adsorption, both equations usually require three further parameters. For the three-plane model these consist of a term for the number of sites "covered"·plus the two constants specified in equations A3.14 and A3.15. For the four-plane model the three parameters consist of a term for the maximum adsorption, a term for the capacitance between the S and the a planes and a binding constant for the adsorbing ion.

The four-plane model is also comprehensive. As indicated above, it can describe in a consistent way the behaviour of a wide range of cations and anions. In both cases, the effects of changing the pH are due to the interplay of the effects of electric potential and the effects on pH. It also has the same ability as the three-plane model to reproduce electrophoretic mobility. This is is illustrated further in Chapter A5.

As for realism, the model could be criticised in two grounds. One is that no attempt is made to specify the surface reactions. However such an approach leads to difficulties with ions that occupy an area equal to that of several sites. Further we can only describe average sites. In reality there are many different surface oxygen atoms occurring on planar, edge and corner sites and involved in one-, two-, and three-fold coordination with metal atoms. This would lead to a very large number of possible reactions. Thus a characteristic of the model is that the surface is only represented in a probabilistic sense. Imperfections in the surface mean that ions are not confined to the planes of adsorption — rather, in a time-averaged sense they are mean planes and their distances apart need not be those of molecular dimensions. Thus the parameters of the model cannot be determined by independent means. They must be determined from the observations of adsorption and charge. A further characteristic is that the model does not involve a choice amongst proposed surface species but rather a choice between known species in solution — for example between HPO_4^{2-} and $H_2PO_4^{-}$.

Another ground for criticising the four-plane model is the way the alternative reactions with the sites are handled. In the three-plane model certain of the reactions are mutually exclusive — a site can react either with a chloride ion to form an outer sphere complex, or with an ion such as phosphate to form an inner-sphere complex. In the four-plane model these possibilities are not mutually exclusive. Because the ligand exchange reaction can displace either a water molecule or a hydroxyl ion from the surface, formation of an outer-sphere complex is not precluded. The site remains "vacant" in this respect. However there is one situation in which a site is no longer "vacant" and competition between different ions can be considered to occur. Thus a site that has reacted with an anion as in equation A3.21 has lost part of its freedom to vary its charge — it can gain a proton to become neutral but it has only two possible charge states instead of three. In terms of the model, if a site is occupied by an anion, it is no longer "vacant" to a hydroxide ion, and vice versa. No account is taken of this in the model and this could be considered as a lack of realism. However including it has little effect (Barrow and Bowden 1987). This is easy to understand for ions such as fluoride and phosphate which have their maximum adsorption at low pH. At low pH there is little competition from hydroxide ions and at high pH there is less adsorption and so the reduction in the number of vacant sites has little effect on hydroxide reaction with the surface. Surprisingly however there is also no effect for silicate adsorption which is greatest at high pH (Barrow and Bowden 1987). The reason is that, when appreciable adsorption of an anion occurs, most of the sites acquire a proton in order to partially balance the negative charge and there is, in fact, little adsorption of hydroxide ions.

Summary

Neither the single-plane model, nor the three-plane model, are comprehensive. At best, they work well under restricted conditions. In contrast, the four plane model works well under a range of conditions. Not only does it describe observations closely but it provides a framework for extension to the more-complex problems of soil. Thus, for example, the effects of pH are subdivided into the effects on the ions in solution and the effects on the charge and potential of the surface. The way that this model works is considered further in Chapter A5 and its application to the problems of soil in Chapter A8.

References

Atkinson, R.J., Posner, A.M. and Quirk, J.P. 1967. Adsorption of potential determining ions at the ferric oxide-aqueous electrolyte interface. Journal of Physical Chemistry 71, 550-558.

Barrow, N J., Bowden, J.W., Posner, A.M. and Quirk, J.P. 1980. Describing the effects of electrolyte on adsorption of phosphate by a variable charge surface. Australian Journal of Soil Research 18, 395-404.

Barrow, N.J., Bowden, J.W., Posner, A.M. and Quirk, J.P. 1981. Describing the adsorption of copper, zinc and lead on a variable charge mineral surface. Australian Journal of Soil Research 19, 309-321.

Barrow, N.J. and Bowden, J.W. 1987. A comparison of models for describing the adsorption of anions on a variable charge mineral surface. Journal of Colloid and Interface Science (In Press).

Barrow, N.J., Madrid, L., and Posner, A.M. 1981. A partial model for the rate of adsorption and desorption of phosphate by goethite. Journal of Soil Science 32, 399-407.

Bowden, J.W., Posner, A.M. and Quirk, J.P. 1977. Ionic adsorption on variable charge mineral surfaces. Theoretical charge development and titration curves. Australian Journal of Soil Research 15, 121-136.

Bowden, J.W., Nagarajah, S., Barrow, N.J., Posner, A.M. and Quirk, J.P. 1980. Describing the adsorption of phosphate, citrate and selenite on a variable-charge mineral surface. Australian Journal of Soil Research 18, 49-60.

Davis, J.A. and Leckie, J.O. 1978. Surface ionization and complexation at the oxide/water interface 2. Surface properties of amorphous iron oxyhydroxide and adsorption of metals. Journal of Colloid and Interface Science 67, 90-107.

Davis J.A. and Leckie, J.O. 1980. Surface ionization and complexation at the oxide/water interface 3. Adsorption of anions. Journal of Colloid and Interface Science 74, 32-43.

Gilbert, G.A. and Rideal, E.K. 1944. The combination of fribrous proteins with acids. Proceedings of the Royal Society (London) A 182, 335-344.

Goldberg, S. and Sposito, G. 1984. A chemical model of phosphate adsorption by soils: 1. Reference oxide minerals. Soil Science Society of America Journal 48, 772-778.

Goldberg, S. and Sposito, G. 1985. On the mechanism of specific phosphate adsorption by hydroxylated mineral surfaces: a review. Communication in Soil Science and Plant Analysis 16, 801-821.

Morel, F.M.M., Westall, J.C. and Yeasted, J.G. 1981. Adsorption models: a mathematical analysis in the framework of general equilibrium calculations. In Anderson, M.A. and Rubin, A.J. (Editors). Adsorption of Inorganics at Solid-liquid Interfaces. Ann Arbor Science Publ. Ann Arbor, Michigan.

Pulver, K., Schindler, P.W., Westall, J.C. and Graver, R. 1984. Kinetics and mechanism of dissolution of

Bayerite (γ-Al(OH)$_3$) in HNO$_3$-HF solutions at 298.2° K. Journal of Colloid and Interface Science 101, 554-564.

Sigg, L. and Stumm, W. 1980. The interaction of anions and weak acids with the hydrous goethite (α FeOOH) surface. Colloids and Surfaces 2, 101-117.

Sposito, G. 1983. On the surface complexation model of the oxide—aqueous solution interface. Journal of Colloid and Interface Science 91, 329-340.

Westall, J. and Hohl, H. 1980. A comparison of electrostatic models for the oxide/solution interface. Advances in Colloid and Interface Science 12, 265-294.

White, R.E. 1980. Retention and release of phosphate by soil and by soil constituents. In Tinker, P.B. (Editor). Critical reports on applied chemistry, Vol.2. Soils and Agriculture. Blackwell, Oxford.

Chapter A4

The rate of reaction with oxides

The models discussed in Chapter A3 are equilibrium models. They assume that the equilibrium position of an adsorption reaction has been reached and that no further reaction occurs. When considering the rate of the reaction, neither of these assumptions is appropriate. At least for short periods of reaction, the approach to equilibrium must be considered; and for long periods some second reaction may follow adsorption. Thus a model for the rate of the reaction may have two rate components — one to describe the rate of the adsorption process and one to describe a possible second reaction.

The rate of the adsorption reaction

When an ion reacts with a charged surface the overall reaction may involve a number of steps. We may speculate that the steps might be something like: the approach of the ion to the surface, the displacement of a water molecule from the surface, gain or loss of protons, and the approach (or departure) of an electrolyte ion to balance the change in surface charge. We do not know whether this is the real sequence of steps but we assume that some such sequence occurs and that one of the steps in this sequence is so much slower than the remainder that it becomes the rate-limiting step. The general equation for such a multi-step reaction at a charged surface is given in Chapter B3. To illustrate this equation, the somewhat simpler form appropriate for a specific adsorbate (phosphate) is given here:

$$dc/dt = - k_1 \, \alpha \, \gamma \, c_t \, m_t \, \exp \, (2 \, F\psi_a/RT) + k_2 s_t \qquad \text{A4.1}$$

The symbols in this equation are defined in full in Table B3.1. It suffices to note here that c_t is the concentration of adsorbate in solution, m_t the concentration of vacant sites, and s_t the concentration of occupied sites. In writing A4.1, appropriate values for some of the constants have been used in order to give a simpler equation. Thus, because the divalent phosphate ion is involved, it was assumed that the number of electrons involved is two. It was also assumed: that the rate-determining step occurs only once for each occurrence of the overall reaction; that the rate-determining step precedes the electron transfer step; and that the rate-determining step does not involve electron transfer. The reasons for these assumptions were given in Barrow et al. (1981). The resulting equation is similar to a second-order forward reaction opposed by a first-order back reaction. However the rate of the forward reaction is affected by the electric potential of the surface and is faster when the potential is positive. For phosphate there was apparently no effect of potential on the rate of the back reaction. Note however that these assumptions were made because they were appropriate to the availabe data for phosphate adsorption on goethite. The behaviour of different adsorbates or adsorbants could be different.

Equation A4.1 (or its more-general form, Equation B3.1) serve as central equations from which equations can be developed for either describing the change in concentration as adsorption proceeds if the initial conditions are specified, or the change in adsorption if the concentration is held constant. They also give rise to expressions for the equilibrium given either of these two conditions. The derivations of these equations are given in Chapter B3.

The equation for the rate of change of concentration has been incorporated into the program RATEOX listed in Chapter B. Before considering the output of this program, we will consider the rate of the reaction that follows adsorption.

The rate of the second reaction

All of the models developed so far have been confined to the surface layers and the solution near it. They all assume that the only "defect" in an oxide crystal is that due to the surface. That is, they argue that it is because there is a surface there are unsatisfied bonds and other irregularities. However crystals may have other kinds of defects. These may be due to impurities, but even pure crystals can have interstitial atoms and also lattice vacancies. Proton reaction with an oxide is probably not confined to the surface groups. At low pH, the protons probably penetrate the crystal giving rise to a cloud of interstitial protons with a distribution not unlike that of the cloud of ions that make up the diffuse layer. However, in this case, the diffusion is in the solid state and in response to the gradient of electrochemical potential. That is, diffusion in the solid state differs in some ways from diffusion in liquids or in gases and so the mathematical description of this cloud will differ somewhat from that of the diffuse layer. At high pH there would be an equivalent cloud of lattice vacancies. Thus, in addition to the gradient of electric potential away from the surface and into the solution, there will be a gradient of electric potential into the solid phase. If it could be argued that the intrinsic ratio of interstitial protons to proton vacancies was unity, then the electric potential in the centre of a crystal would be taken as zero. If this ratio were not unity the electric potential in the centre would not be zero and there would be a difference between this potential and that in the bulk solution. The slowness of titrations with goethite and the apparent hysteresis effects would be explained by the time required for protons to diffuse into, and out of, the crystal.

When metal ions, such as those of nickel and zinc react with goethite, there also appears to be a continuing reaction (Chapter A2). This may also be due to diffusion of foreign ions into the lattice, partly as interstitial ions, partly by occupying vacancies in the iron lattice, and partly by displacing iron. It is of interest that nickel seemed to be more favoured in this respect than zinc or cadmium. Again this diffusion would give rise to a cloud of "foreign" ions and a gradient of electric potential into the surface.

For anions, such as phosphate, reported values of the rates are lower and diffusion of anions in crystals may be restricted to bigger defects. It would therefore be interesting to observe the rate of phosphate reaction with goethite crystals that have been "healed" using the heat treatment method of Schwertmann et al. (1985).

There are two problems in developing a model against which to test the hypothesis that the continuing reaction is due to diffusion of adsorbate into the crystal. The first is that diffusion in the solid state is along a gradient of electrochemical potential. Part of this gradient is therefore the gradient of electric potential into the crystal. As described above, theories do not seem to have yet been developed to describe this gradient. There is no fully satisfying way around this problem at present. All that can be done is to model diffusion along the gradient of chemical potential towards the interior of the particle and to note that the model will need to be modified when appropriate theories are developed to describe the gradient of electric potential. The second problem is related. If diffusion occurs, charge will be transferred from the plane of adsorption to the particle itself. This charge will affect the electric potential of the s plane and this will, in turn, affect the electric potential of the a

31

plane. That is, there will be a feed-back effect. Again a satisfactory model requires a detailed understanding of how this charge is distributed and of its effect on the potential. In order to circumvent the impasse produced by this lack of detailed theory, two assumptions were made. These are described in the next section.

Assumptions in developing a model for the rate of the second reaction

The first assumption is similar to that of Barrow (1983) and is that diffusion could be described by:

$$M = 2\,C_{is}\,\sqrt{(\tilde{D}\,f\,t/\pi)} \qquad\qquad A4.2$$

where M is the amount of material diffusing into a particle during the period t while the surface concentration of the ion i is constant at C_{is}, and $\tilde{D}$ is a diffusion coefficient. The term f is known as the thermodynamic factor and is derived from the ratio of the log. of surface activity to the log. of surface concentration.

There are two aspects of this equation that merit attention at this stage. One is that it is the equation for diffusion into a plane. That is, it is implied that the rate of diffusion is so slow that we can ignore the shape of the particles and thus ignore the "radial convergence" of diffusion lines. The other aspect is that the term C_{is} refers to surface concentration and its dimensions are therefore moles per area rather than moles per volume. This means that the dimensions of $\tilde{D}$ are time $^{-1}$. $\tilde{D}$ is related to the normal volumetric diffusion coefficient through the "thickness" of the surface layer. Equation A4.2 can only hold while C_{is} is constant.

The second assumption is that the charge on the diffusing ion can simply be added to the charge in the s plane in order to calculate the new value for the electric potential in the s plane. Before making this assumption, the effect of locating the charge in a further plane of electric potential inside the s plane was investigated (Barrow 1986). It was found that the location of this plane had no effect on the potential in the s plane. That is, the potential in the s plane was affected by the amount of charge but not by the value allocated to the capacitance between the s plane and the inner plane. It was therefore simpler to make the planes coincident and to add the charge to the s plane. This changes the electric potential of the s plane and also changes the electric potential of the a plane. Consequently, the concentration of adsorbed ions in the a plane (C_{is}) will change even though the solution concentration be kept constant. The process is then followed using a series of small steps and it is assumed that during each step C_{is} remains constant. To illustrate the required modification of Equation A4.2, let the initial surface concentration be C_0, and let this change at t_1 to C_1. Then at a subsequent time t, a linear superposition of equations like A4.2 gives:

$$M = (2/\sqrt{\pi})\,(C_0\,\sqrt{(\tilde{D}\,f\,t)} + (C_1 - C_0)\,\sqrt{(\tilde{D}\,f\,(t-t_1))}) \qquad\qquad A4.3$$

Description of data using the model for rate of reaction

The model was applied to the rate of reaction of phosphate with a sample of goethite as described by Madrid and Posner (1979). Fig. A4.1 shows that the reaction was much faster

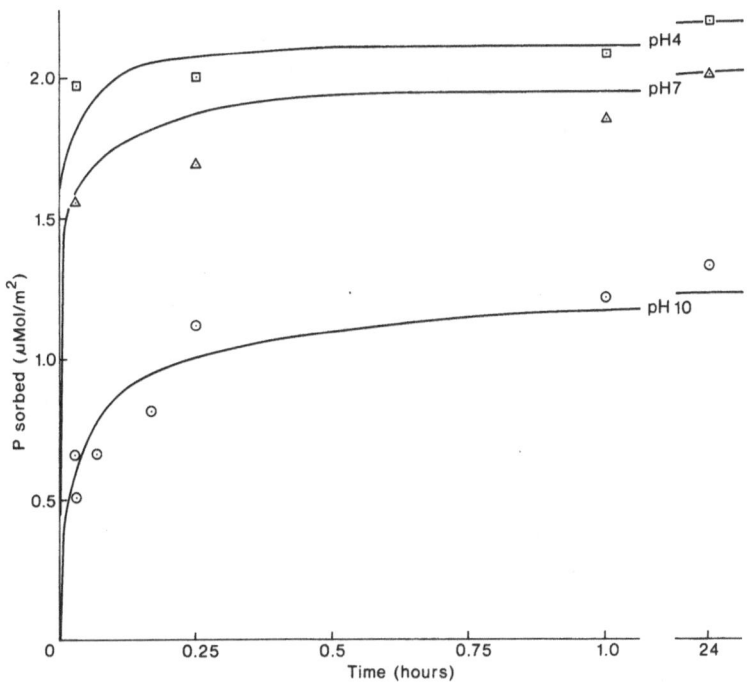

Fig. A4.1   Modelling the rate of reactioon of phosphate and geothite over short periods. The data were obtained from Fig. 9 of Madrid and Posner (1979). The lines are from the model described in this Chapter.

at pH 4 than at pH 10. These differences in the initial rate of reaction were modelled using equations derived from Equation A4.1. This equation shows that the rate of the forward reaction is faster when the electric potential is large — that is at low pH. The initial adsorption reaction was modelled as being completed in periods as short as a few minutes at pH 4. Further reaction is then due to the slow diffusion reaction. This is not well demonstrated in Fig. A4.1 because the data did not extend to long periods. Madrid and Posner (1979) were better able to demonstrate the slow reaction by adding smaller amounts of phosphate. A relatively small increase in sorption then produced a relatively large decrease in the solution concentration. Fig. A4.2 shows that the model reproduced the changes in concentration with time fairly well — the main discrepancy was due to the difficulty of reproducing the correct amount of curvature in the concentration direction at these low values for concentration in solution.

    The rates of the continuing reaction reported by Madrid and Posner (1979) for phosphate were low. A much more pronounced continuing reaction was reported by Gerth and Bruemmer (1981, 1983) for the reaction of zinc, cadmium and nickel with goethite. Fig. A4.3 shows that the rate effects were in the sequence nickel > zinc >

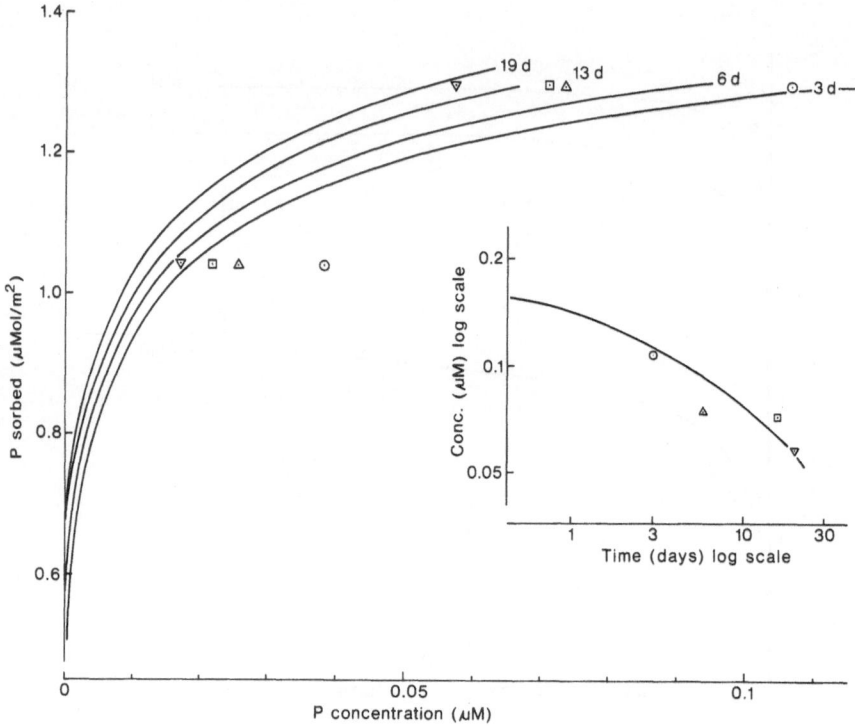

Fig. A4.2.   Modelling the rate of reaction of phospate and geothite over long periods. The data were obtained from Fig. 7 of Madrid and Posner (1979). The lines are from the model.

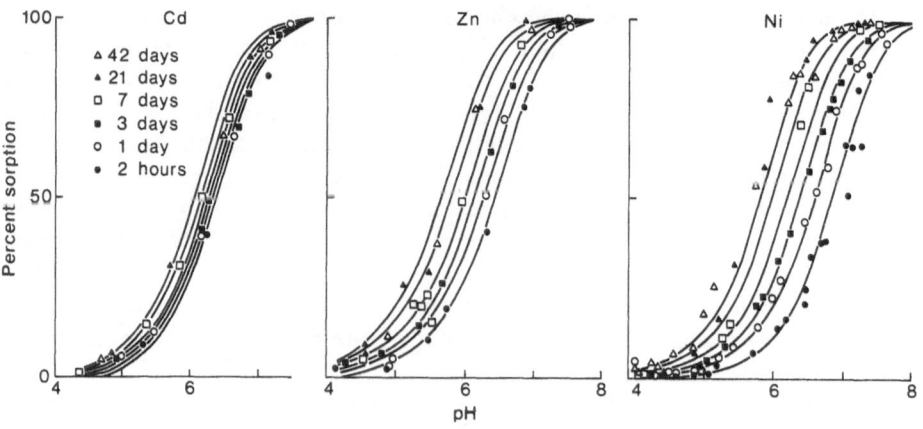

Fig. A4.3   The effect of time and pH on the sorption of Cd, Zn and Ni by geothite. The lines were derived from a model that permits diffusion and metal ions into the crystal. Details will be published subsequently.

34

cadmium and that the model was able to reproduce them. Fitting the model to these data was done with the cooperation of Professor Bruemmer. It is intended to publish a description of the application of the model to these results jointly with him. In order to avoid prejudicing this publication, details will not be given here. The goethite used by Gerth and Bruemmer (1981, 1983) had been prepared by permitting the crystals to grow at 60° for two days. This is somewhat shorter than the six days used by Schwertmann *et al.* (1985) and this raises the possibility that crystal development was not complete and crystallinity was not high. Thus we do not know whether the marked continuing reaction observed was simply a property of the metal ions or instead a property of the combination of the ions and the particular preparation of goethite.

Consequences of the model of the rate of reaction

It was noted in Chapter A2 that soil goethites often contain large amounts of silicon and phosphate. It is assumed that this will make their surface charge more negative. Further, soil goethites may occur in close association with clay minerals and therefore in close association with permanent negative charge. In this section we explore the possible consequences of this negative charge. As explained above, it is assumed that this negative charge effectively resides in the s plane. Fig. A4.4 shows that as the amount of this charge

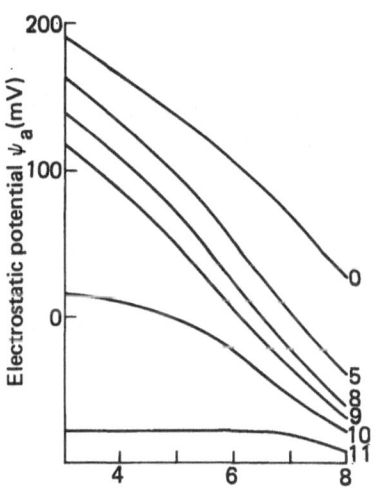

Fig. A4.4   Effect of adding charge to the s plane on the electrostatic potential in the a plane. Numbers on the curves indicate the added negative charge (μmol M^{-2}).

increases, the electric potential in the a plane decreases. However the effects are fairly complex. At low pH, they become large as the amount of charge approaches, and exceeds, the maximum adsorption of protons in the a plane. In addition, the slope in the pH direction decreases to zero. This occurs because, at moderate pH, the s plane becomes saturated with protons balancing the negative charge. Decreases in pH cannot then produce any increase in the number of adsorbed protons. We then have the paradoxical situation of a "variable-charge" material for which the charge does not vary with pH.

In Chapter A8 it will be suggested that the electric potential of the variable charge materials in soils varies between particles. Let us consider the possibility that this might arise as a result of different amounts of negative charge in close association with the particles. Thus Fig. A4.4 is a model (albeit a crude one) of this situation. Note that the electric potential of the most-negative sites is not affected by pH. These are the sites most likely to react with cations.

Summary

This brief chapter outlines the problems of modelling the rate of reaction between oxides and ions. It describes a well-based model for the rate of the initial adsorption reaction. It then discusses the problems of modelling the slow diffusion penetration of the surface that follows the initial adsorption reaction. Appropriate theory does not seem to be available for handling the gradient of electric potential that this diffusion produces. Nevertheless an approximate model is described and it is shown that it describes fairly well the slow reaction of phosphate, cadmium, zinc and nickel with goethite.

References

Barrow, N.J., Madrid, L. and Posner, A.M. 1981. A partial model for the rate of adsorption and desorption of phosphate by goethite. Journal of Soil Science 32, 389-407.

Barrow, N.J. 1983. A mechanistic model for describing the sorption and desorption of phosphate by soil. Journal of Soil Science 34, 733-750.

Barrow, N.J. 1986. Testing a mechanistic model. VI. Molecular modelling of the effects of pH on phosphate and on zinc retention by soils. Journal of Soil Science 37, 000-000.

Gerth, J. and Bruemmer, G. 1981. Einfluss von Temperatur und Reaktionzeit auf die Adsorption von Nickel, Zink, und Cadmium durch Goethite. Mitteilungen der Deutschen Bodenkundlichen Gesselschaft 32, 229-238.

Gerth, J. and Bruemmer, G. 1983. Adsorption und Festlegung von Nickel, Zink und Cadmium durch Goethite (αFeOOH). Fresenius Zeitschrift fuer Analytical Chemistry 316, 616-620.

Madrid, L. and Posner, A.M. 1979. Desorption of phosphate from goethite. Journal of Soil Science 30, 697-707.

Schwertmann, U. Cambier, P. and Murad, E. 1985. Properties of goethites of varying crystallinity. Clays and Clay minerals 33, 369-378.

Chapter A5

The four-plane model and how it works

¹Chapter A3 compares the three models that have been used to describe the adsorption of anions and cations on variable-charge surfaces. It shows that the four-plane model is the best. The purpose of this chapter is to help the reader become more familiar with the model.

There are several ways we can approach a model such as this. One is by a formal description of the assumptions involved as in Chapter A3. Another is by detailed presentation of the equations as in Chapter B2. These are, of course, essential; but they don't really enable us to get to grips with the model. The best way to do this is to "twiddle the knobs" — to change the values of the numbers and to observe the effect. This is why the computer programs are made available — you can do it yourself. However, to get you started, this chapter will present some results of knob-twisting.

It is convenient to think of the numbers associated with the model as occurring in three categories. One of these comprises the parameters. These are intended to reflect the properties of the system. They include the binding constants for the various ions, the capacitances between the mean planes, and the adsorption maxima. We can change the values of these parameters in the model and observe the effects on the second category of numbers — the dependent variables. The dependent variables include the charges on the various planes and also the electric potentials. In a given experiment, some (but not all) of these dependent variables will have been measured. The third category of numbers is the independent variables. These include the pH, the ionic strength of the solution, the concentration of the ions, their charge and their valency. The effects of varying some of these variables on some of the dependent variables can be observed. Thus, in these "knob-twisting" exercises we can choose to observe the effect on the model of varying the values of the parameters and also of varying the values of the independent variables.

Varying the charge parameters

Consider first the parameters that describe the charge. These comprise: the maximum adsorption in the s plane; the binding constants for H^+, OH^-, cations and anions; and the capacitances between the three planes that are relevant in the absence of specifically adsorbed ions.

The effects of varying these parameters are illustrated in Fig.A5.1. In each case, the solid line shows the output when the parameter values suggested in the program listing are used. These are not the "correct" values — merely values that may describe the behaviour of goethite fairly well.

As might be expected, increasing the value of the maximum adsorption in the s plane increases the modelled value for charge (Fig.A5.1a). The effects are however not large. Thus the titration curve does not provide a sensitive means for fixing the value for this parameter. The modelled value for charge is also increased when the capacitance between the s and the β planes is increased (Fig.A5.1b). A large value for this capacitance implies

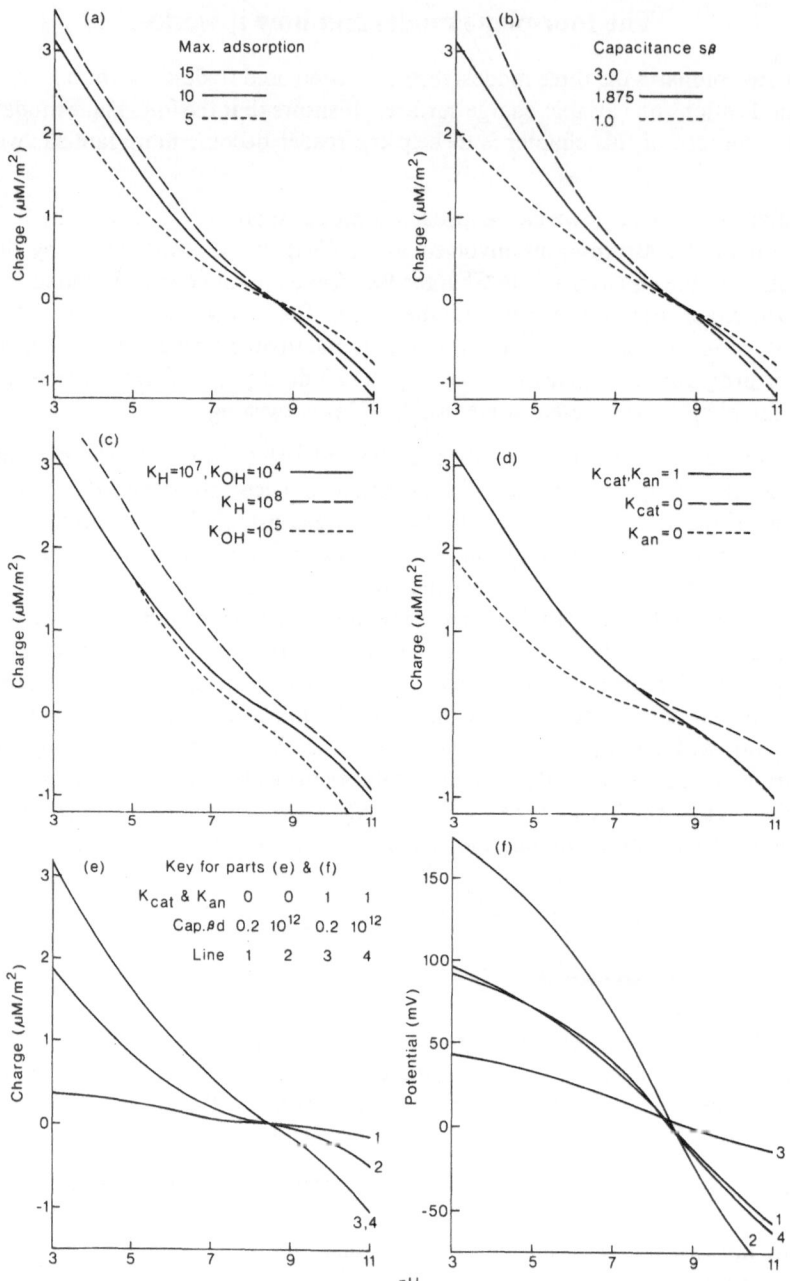

Fig. A5.1   Exploring the effects of varying the parameters of the four-plane model on the modelled values for charge. In each case the solid line represents the output when the suggested values indicated in the listing of program "Bowden" are used. The effects of variations from those values are shown. The parameters varied are: a) the maximum adsorption in the s plane; b) the capacitance between the s and the β planes; c) the value for the constants K_H and K_{OH}; d) the values for K_{cat} and K_{an}; e) and f) the values for the capacitance between the β and the d planes and the constants K_{cat} and K_{an}. Part f differs in showing the modelled values for the electric potential in the d plane.

that the β plane is located close to the s plane. Consequently, at low pH, the closer approach of the electrolyte anions tends to decrease the electric potential in the s plane. This is opposed by an increase in the number of protons and thus a higher charge. Similarly, at high pH and negative charge, the closer approach of electrolyte cations leads to an increase in the negative charge. When only one ionic strength is considered, as in Fig. A5.1, the effects of varying the maximum adsorption and the effects of varying the capacitance are somewhat similar. On such limited data it would be possible to largely compensate for a high value of one by a low value of the other. However the effects of these parameters differ when ionic strength is also varied. Such data are essential if a realistic choice of the values of these parameters is required.

The relative values for the K_H and the K_{OH} parameters affect the point of zero charge. Thus, increasing the value for K_H by one unit increases the point of zero charge by 0.5 pH units (Fig. A5.1c). Similarly increasing K_{OH} by one unit decreases the point of zero charge by 0.5 units. The effects of varying K_H show mainly below the point of zero charge because this is the range in which proton adsorption dominates; the effects of varying K_{OH} show mainly above the point of zero charge because this is the range in which hydroxyl adsorption is modelled as occurring. However if we increase *both* K_H and K_{OH}, the point of zero charge is unchanged but the slope of the line relating charge and pH is increased. The slope of this line is also affected by the values allocated to the parameters K_{cat} and K_{an} (Fig. A5.1d). When the value of K_{an} is zero, the concentration of anions in the β plane is also zero. This means that, at low pH, fewer protons can enter the s plane and so the charge is lower. Similarly, when K_{cat} is set to zero the negative charge at the high pH is small. The values of these two parameters interact with the value allocated to the capacitance between the β and d planes. When an appreciable value is allocated to K_{an} and K_{cat}, the value of this capacitance has no effect on the charge (lines 3 and 4 of Fig. A5.1e). There are however effects when K_{an} and K_{cat} are set to zero. In this case a large value for the capacitance gives a higher charge because the ions of the diffuse layer are then located closer to the s plane.

The interaction between K_{an}, K_{cat} and the capacitance β-d is especially relevant if it is desired to also model the zeta potential. This is the potential of the slip plane when the particles are subjected to electrophoresis. It is sometimes assumed that the zeta potential can be equated with the potential in the d plane. This assumption is controversial but most agree that the potential in the d plane must be close to the zeta potential. Values for the potential in the d plane are shown in Fig. A5.1f for the various values of K_{cat}, K_{an} and capacitance β-d. Values for the potential are largest at low pH when K_{an} and K_{cat} are zero and when the capacitance β-d is large. They are decreased both by increasing the values of K_{an} and K_{cat} and by decreasing the value of the capacitance. The resulting curve (line 3 of Fig. A5.1f) is close to values usually observed. That is, appreciable values for K_{an} and K_{cat} and a small value for the capacitance β-d are needed to model the zeta potentials.

This "knob-twisting" exercise shows that, if the data available is limited, the model is "over-parameterised" — and the same charge curve could be produced by a wide variety of combinations of values. For some purposes this could be regarded as convenient because it means that the model can easily be made to fit data! Nevertheless, permitted values of the parameters can be limited if more than one kind of data are available. For example, the parameters K_{an}, K_{cat} and capacitance β-d are sensitive to the values for the zeta potential. Thus, the values of these parameters could be best determined if values for this potential were available. In the absence of such data, however, it was sometimes convenient to allocate a large value to the capacitance β-d as this made the solution of the simultaneous equation easier. The method now provided is more robust and the only penalty for using a small value for the capacitance is that rather more iterations are needed. A value for the maximum adsorption in the s plane cannot be accurately determined from data for charge. One way to estimate it is to measure tritium exchange. In this model, a "site" is defined as having an average of three protons at the point of zero charge. Thus the number of sites is one third of the exchangeable protons. It is for this reason that a value of $10\ \mu$ moles m^{-2} is suggested as an approximate value for the maximum number of sites. Finally, the value for the capacitance s-β is sensitive to the effects of ionic strength and data for charge measured at several ionic strengths are needed in order to allocate a reliable value to this parameter.

Varying the parameters that describe adsorption

There are three additional parameters that determine the characteristics of adsorption. These additional parameters are, of course, not required to model charge in the absence of adsorption. These three parameters are: the maximum adsorption in the a plane; the binding constant for the adsorbing ion; and the capacitance between the s and the a planes. The effects of varying the value for the maximum adsorption are fairly obvious and are not illustrated. This parameter is required because the observed maximum adsorption of anions differs widely. Thus, for fluoride, it is about four times that of phosphate. Such differences reflect differences in ion size and in orientation on the surface. The effects of varying the magnitude of the binding constant are also fairly obvious. This term occurs in equations as a product with concentration. Therefore increasing the magnitude of the binding constant simply means that a given effect is reached at a lower concentration. In contrast, the effects of varying the capacitance between the s and the a planes are somewhat more complex.

It is convenient to assume, for the purposes of illustration, that the distance between the s plane and the a plane is inversely proportional to the capacitance s-a. Thus Fig.A5.2a can be interpreted as showing how the electric potential decreases with distance from the s plane. The model permits the a plane to be located anywhere between the s and the β planes according to the value allocated to the capacitance s-a. In Fig.A5.2a the a plane is shown at the position appropriate to the value given in the program listing for phosphate. It is apparent from Fig.A5.2a that a molecule sited between the s and the β planes would experience a different change in potential with pH depending on how close it was to the s plane. This is illustrated in Fig.A5.2b for a molecule sited in three of the many possible positions — that is: in the s plane; in the indicated position of the a plane; and in the β plane. The figure shows that the change in electric potential with pH is greatest for the molecule sited in the s plane. This has consequences for adsorption. For anions, if the adsorbing molecules were located in the s plane, there would be large adsorption at low pH, because of the high electric potential. Adsorption would however fall steeply with increasing pH because of the large change in potential (Fig. A5.2c). At the other extreme, if the adsorbing anions were located in the β plane, the change in potential would be too small to counteract the increasing concentration of adsorbing ion (the output is for a divalent ion with pK$_2$ of 7). Thus for anions, the location of the adsorbing molecules, as reflected by the capacitance s-a, has a large influence on the effect of pH on adsorption.

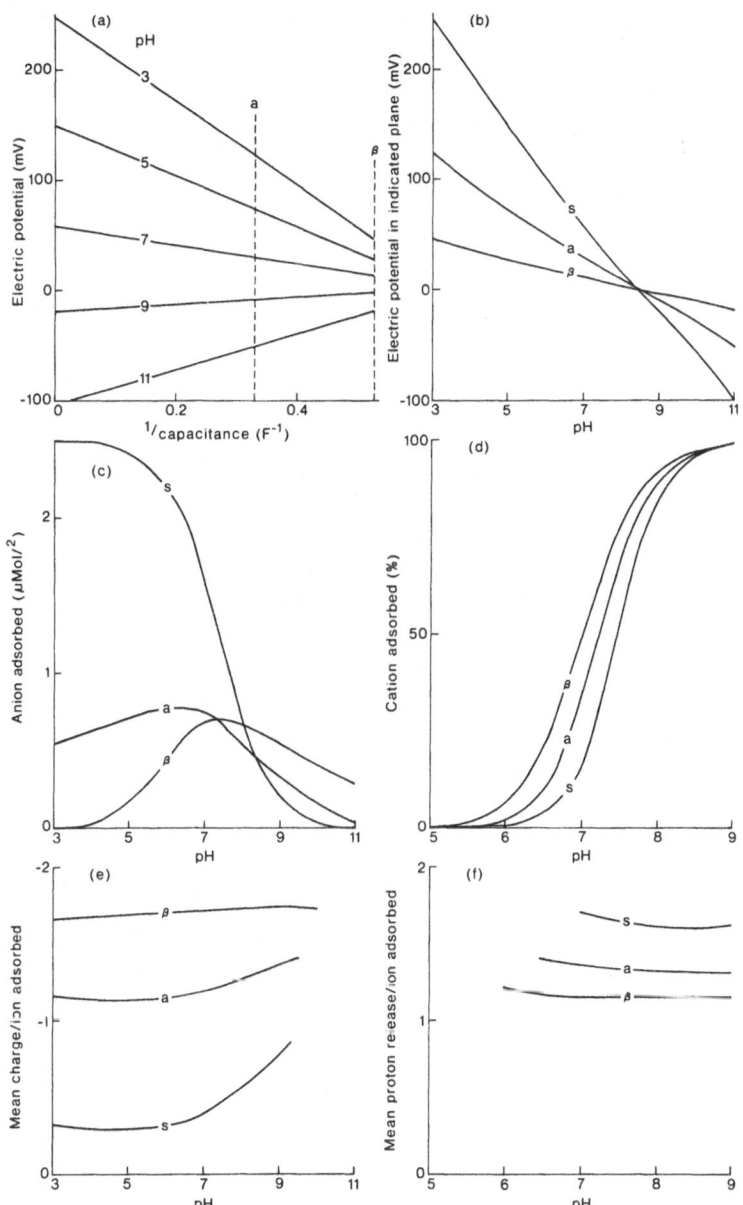

Fig A5.2   Exploring the effects of varying the capacitance between the s and the a planes on modelled adsorption, and charge balance, for anions and for cations. The other parameters were held constant at the suggested values indicated in the listing of programs "Bowden" and "Metals". The outputs are for adsorption of a hypothetical divalent anion with pK_2 of 7 or for a hypothetical monovalent cation MeOH$^+$) with pK_1 of 7. Both were in a 0.1M concentration of a 1:1 electrolyte. The product of the binding constant and the concentration of the adsorbing anion was 10^6 and the initial concentration of metal was 100 μ moles per litre. Part a shows the modelled values for electric potential plotted against the reciprocal of capacitance for indicated values of pH. Part b shows the modelled values for electric potential plotted against the reciprocal of capacitance for indicated values of pH. Part b shows the change in the electric potential with pH in the planes identified in part a as S, a or β. The remaining parts show the effects of pH on adsorption and on charge for adsorbing ions located in these planes.

For cation adsorption, analogous effects occur but are not as large. Thus if the adsorbing cations were located in the s plane, the higher values for potential mean that adsorption is delayed until a higher pH is reached but the subsequent change with pH is steeper (Fig. A5.2d). The less-extreme effects on cation adsorption mean that cation adsorption does not provide as sensitive a test of the model as anion adsorption.

The value allocated to the capacitance between the s and the a planes also affects the way the charge on the adsorbed ions is balanced. If the value for the capacitance is large so that the adsorbed ions are very close to the s plane, the charge on the adsorbed ions tends to be balanced in the s plane — that is, for an adsorbing anion, hydroxyls are displaced from, or protons enter, the s plane. In this case, the mean charge conveyed to the adsorbing surface is small. If the value for the capacitance is smaller, so that the ions are located closer to the β plane, the charge tends to be balanced by displacement of electrolyte anions from, or increased adsorption of electrolyte cations in, this plane. In this case the mean charge conveyed to the adsorbing surface is larger. These effects are illustrated in Fig. A5.2e for three possible positions of the adsorbing ion. For an adsorbing cation a similar argument holds but, of course, with opposite charges. In this case, however, it is conventional to report the protons released rather than the charge conveyed to the surface. This value is largest when the value for the capacitance is large so that the adsorbing cations are located very close to the s plane (Fig. A5.2f). Note that, in this figure, the values modelled are all greater than one. This is because the model postulates that the monovalent metal ions are adsorbed. One proton therefore comes from the hydrolysis of a divalent ion to "replace" the adsorbed monovalent one; the other (part of a) proton comes by displacement from the surface. Especially for anion adsorption, measured values for the charge conveyed to the surface provide a sensitive test for the value of the capacitance term and thus a robust test of the model.

Varying the dissociation characteristics

Amongst the important variables that can be controlled by the experimenter are the dissociation characteristics of the adsorbate. An experimenter is, however, confined to those values that are available in nature. A modeller has the advantage that he may choose any value he wishes in order to explore the characteristics of the model. Thus in Fig. A5.3a, the adsorption of four hypothetical monovalent anions is modelled. At low pH, there are large differences between the anions because only those derived from acids with low values for the pK are sufficiently dissociated to provide an adequate concentration of the monovalent ion. For the species with a pK_1 of 3, adsorption decreases with increasing pH. This is because, by pH 3, it is already half dissociated and the adsorption equation is dominated by the decreasing electric potential. For the other species the adsorption increases at first with increasing pH because there is a 10-fold increase in the concentration of the monovalent ion with each unit increase in pH. This more-than counter balances the decreasing electric potential. (The effects of pH on electric potential are affected by the value allocated to the capacitance s-a — as discussed earlier.) At high pH the adsorption of all the species approaches a common line because all species are fully dissociated and the only effect of pH is that due to the electric potential.

The adsorption of the hypothetical cations differs in that the effects of potential and of dissociation are complementary rather than opposing. Increasing the pH causes

42

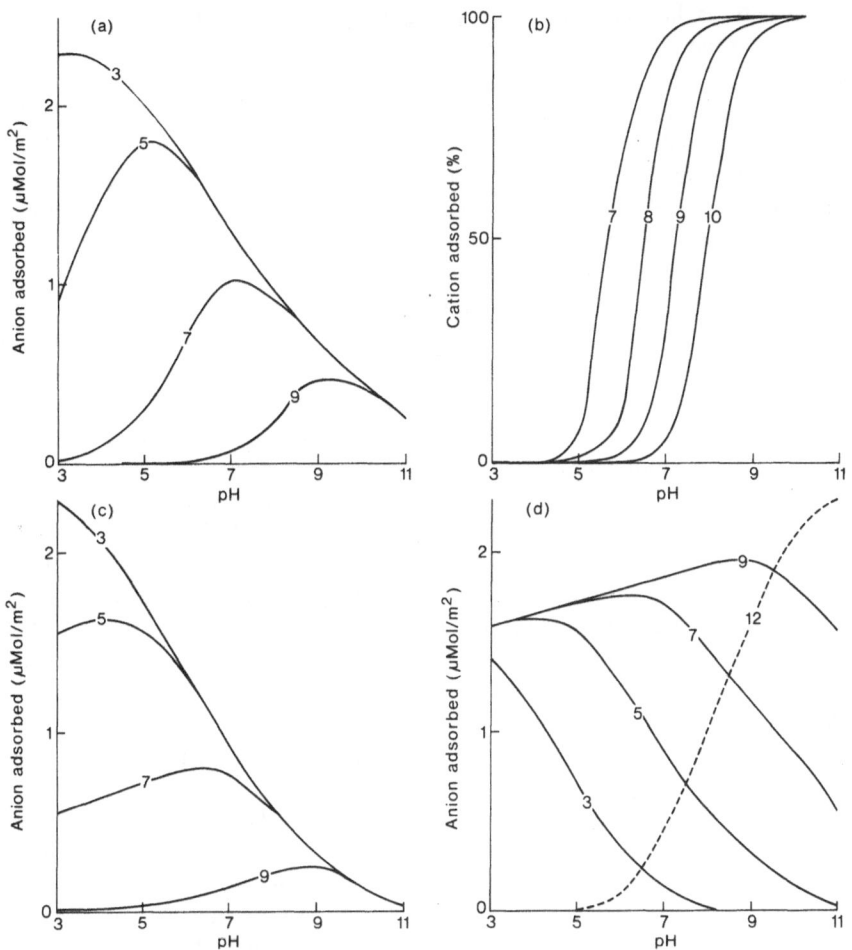

Fig. A5.3 Exploring the effects of varying the dissociation characteristics of the adsorbate. Conditions were similar to those of Fig. A5.2. Part a) shows the effect of varying the pK$_1$ when a monovalent anion is adsorbing; part b) shows the effects when a monovalent cation is absorbing c) shows the effects of varying the pK$_2$ when a divalent anion is adsorbing; part d) differs from part c) in that the binding constant is also varied so that the product of binding constant and dissociation constant did not vary. The dotted line in the figure shows the output when the pK$_2$ was 12 but the pK$_1$ was 10. For all the other anions the pK$_1$ was set at 0.

dissociation, as for anions, but for cations the decreasing electric potential is favourable rather than unfavourable. The result is a fairly sudden increase in adsorption with increasing pH (Fig. A5.3b). The lower the value for pK, the lower the pH at which adsorption begins. This is because an adequate value for the product of the concentration term and the term for electric potential is reached at a lower pH. Note that adsorption begins well below the pK value of the hypothetical cation and thus at a pH at which the concentration of monovalent ions would be small. The pH at which adsorption begins can be changed, within the model, by changing the value assigned to the binding constant for monovalent cations. Adsorption can begin at even lower pH values if the binding constants are made larger. Thus, in the model, it is not merely the concentration of

43

monovalent cations that determines adsorption; it is the product of the concentration, the binding constant, and the term for electric potential.

Adsorption of divalent anions is broadly similar to that for the monovalent anions. The difference is that, because the ions are divalent, the electric potential of the surface has a larger effect. As a result, at low pH, and positive surfaces, adsorption is greater than for monovalent ions (Fig. A5.3c). Furthermore, as the pH is increased, the decrease in the potential has a greater effect so that the peaks in adsorption which occur close to the pKs are not as well marked. Finally, lower adsorption is recorded at high pH.

In considering the effects of varying the pK it has been assumed in the presentation so far that this is the only parameter to be altered. However, it is found that, for anions, there is a good correlation between the pK for dissociation of the relevant acid and the log. of the binding constant — the higher the pK the higher the log. of the binding constant (Barrow and Bowden 1987). That is, ions that have a strong affinity for protons also have a strong affinity for the electrophilic metal ions of the oxide surface. To illustrate this effect, the product of the dissociation constant and the binding constant has been kept constant in Fig. A5.3d. That is, when the pK was large (and the dissociation constant small), a large value was allocated to the binding constant. The figure shows that a common value for adsorption is now approached at low pH. Adsorption increases with increasing pH because of the increasing concentration of the adsorbing ion but there is an inflexion near the pK. To add a further complexity, consider the case in which the pK_1 and the pK_2 for dissociation are close. (In the presentation so far it has been assumed that they are far apart.) However if they are close, then the influences overlap, and the concentration of the divalent ion increases steeply with increasing pH. This is the case for silicate (Fig. A1.1). This effect is also illustrated in Fig. 5.3d and it is shown that sorption now increases steeply with increasing pH.

These manipulations of the values for the pK have been deliberately kept hypothetical in order to use them to show how the model behaves. However the comparisons with real behaviour, as outlined in Chapter A3, will be apparent. Thus, in Figure A5.3a, parallels may be found for fluoride for which the pK_1 is near 3 and for borate with pK_1 near 9. In Fig. A5.3b there are parallels with metals such as copper, zinc and cadmium. In figures A5.3c,d there are parallels for sulfate and selenate for which the pK_2 is even lower than the lowest shown, for phosphate with pK_2 near 7, and for selenite with pK_2 near 8.

Varying the background electrolyte

Another variable that can be controlled by the experimenter is the concentration of the background electrolyte. This has often been utilized in soil studies and there are many reports of the effects of electrolyte concentration on sorption of ions by soils. The way that these effects are modelled is illustrated in Fig.A5.4. Part a of this figure shows that increasing the concentration of salt decreases the absolute value of the electric potential in the surface region. Below the point of zero charge, and thus with positively charged surfaces, one can envisage a greater concentration of electrolyte anions close to the surface leading to a lower potential. Above the point of zero charge the analogous effect is obtained with electrolyte cations. These lower absolute values of the potential permit greater retention of protons (at low pH) or hydroxyls (at high pH) and so greater values for

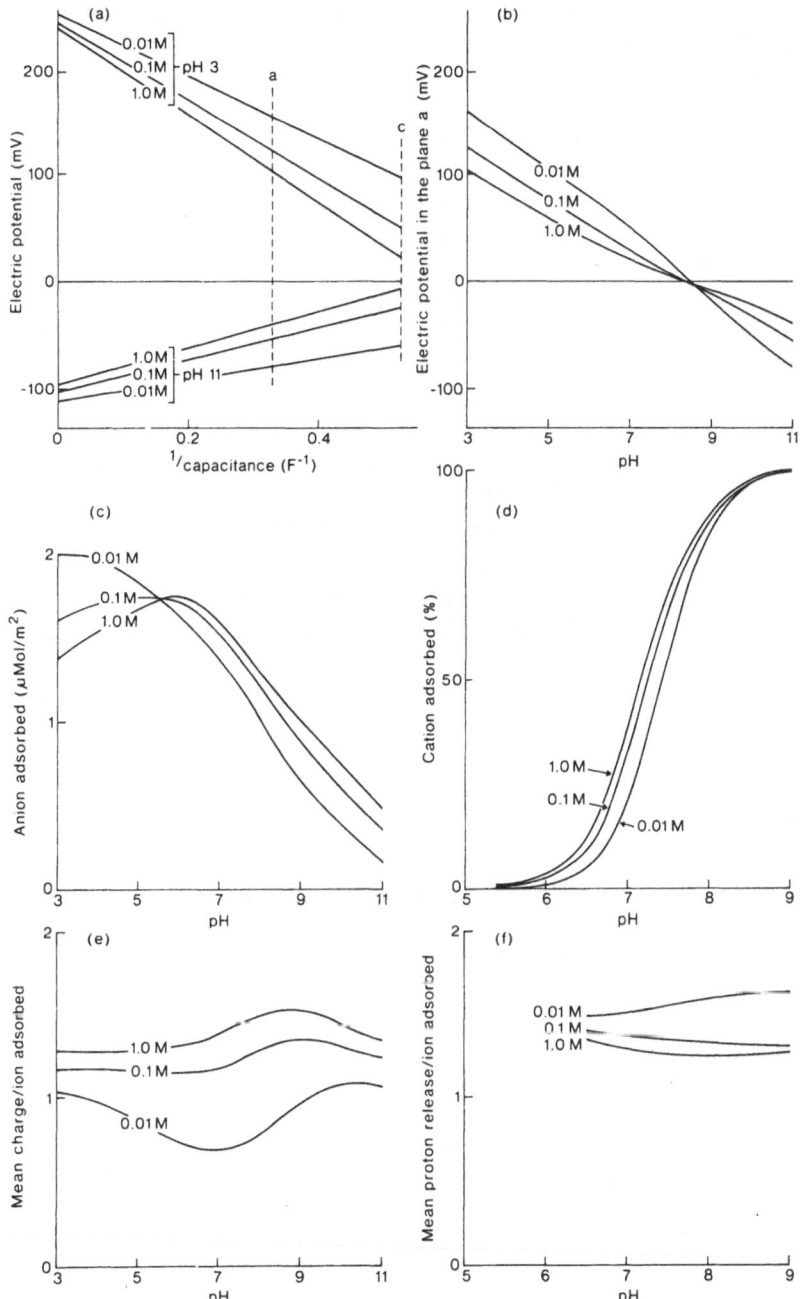

Fig. A5.4 Exploring the effect of varying the concentration of 1:1 electrolyte on modelled adsorption and charge balance. Part a) shows the modelled values for the electric potential plotted against the reciprocal of capacitance for three concentrations of electrolyte. Part b) shows the changes with pH in the electric potential for an ion located in the indicated position of the plane a for three different electrolyte concentrations. The remaining parts show the effects of electrolyte concentration and pH on adsorption and on charge for anions or cations located in the indicated position of the plane a.

the absolute charge. Thus the combinations are: more salt; lower (absolute) potential; more (absolute) charge. Because of the changes in the values for potential are in opposite directions on either side of the point of zero charge, the slope of plots of potential versus pH will have different values depending on the salt concentration (Fig. A5.4b). This, in turn, means that the slope of plots of ion adsorption versus pH will be affected by the electrolyte concentration. Thus, for anions, at low pH, adsorption is greatest in dilute solutions because the positive potential is decreased by increasing the electrolyte concentration. The point at which no effect of salt occurs is at zero potential. This is at a lower pH than the original point of zero charge of the oxide because of the effects of anion adsorption in decreasing the surface charge. These modelled effects are similar to those observed with goethite and it has been shown that the model can be fitted so as to closely describe these observations (Barrow *et al.* 1980). They have very important implications in interpreting the observed effects of electrolyte concentration on anion adsorption by soil. If increasing the concentration of an indifferent electrolyte is observed to decrease anion adsorption, it is reasonable to infer that the adsorbing surfaces have, on the average, a positive potential. Similarly, if adsorption increases, it is reasonable to infer that the adsorbing surfaces have, on the average, a negative potential.

For cation adsorption the model also predicts effects of electrolyte concentration (Fig. A5.4d). Because increasing the electrolyte concentration decreases the potential at pH values below the point of zero charge, it increases the modelled value for adsorption. No point of zero salt effect is seen in this figure because it would occur at a pH above 8. If, however, the electrolyte were to contain a cation with appreciable value for K_{cat}, the increased amount of the cation in the β plane would tend to decrease adsorption. That is, the effect would be in the opposite direction. A further complication is that many anions form complexes with metal ions in solution. If the electrolyte contains such anions, the effect of increasing the electrolyte concentration will depend on the affinity of the anion for the metal and on the affinity of the complex for the surface.

Increasing electrolyte concentration also has a large effect on the modelled values for the charge on the surface when ions adsorb. With a concentrated electrolyte, much of the charge on the anions adsorbed will be balanced by re-adjustment of the electrolyte ions near the surface — at low pH by decreases in the amounts of electrolyte anion, at high pH by increases in the amounts of electrolyte cation. Thus the charge conveyed to the surface by an adsorbing divalent anion is recorded as large (Fig. A5.4e). With a dilute electrolyte, rather more of the charge balance is made up by adjustments in the proton (and hydroxyl) ions on the surface — at low pH rather more protons would be adsorbed, at high pH rather fewer hydroxyls. The charge conveyed to the surface is then recorded as lower. Furthermore at low electrolyte concentration there are interactions with pH — that is, the way the charge is balanced also varies with pH. At low pH there is a strong tendency to displace electrolyte anions; at high pH there is a strong tendency to attract electrolyte cations. In both cases the charge conveyed to the surface is larger than at intermediate pH values. At these intermediate values, the charge on the surface is small. There is therefore less scope for changes in the electrolyte ions. The charge is therefore balanced by changes in the protons and hydroxyl ions and the change in charge due to anion adsorption is recorded as passing through a minimum.

There are similar effects of electrolyte concentration on the charge balance when cations

adsorb. These are shown in Fig. A5.4d in terms of the protons released. All values are ·
recorded as falling between 1 and 2 because of the proton released when a divalent cation
in solution dissociates to "replace" an adsorbed monovalent cation.

Varying the amount of adsorption

The final variable considered here is the amount of adsorption. This is under the control of
the experimenter who can vary the amount of adsorbate added to the system. With
increasing amounts of anion adsorption, the modelled values for the surface charge
decrease (Fig. A5.5a). The lines are however not straight. They tend to pass through a zone

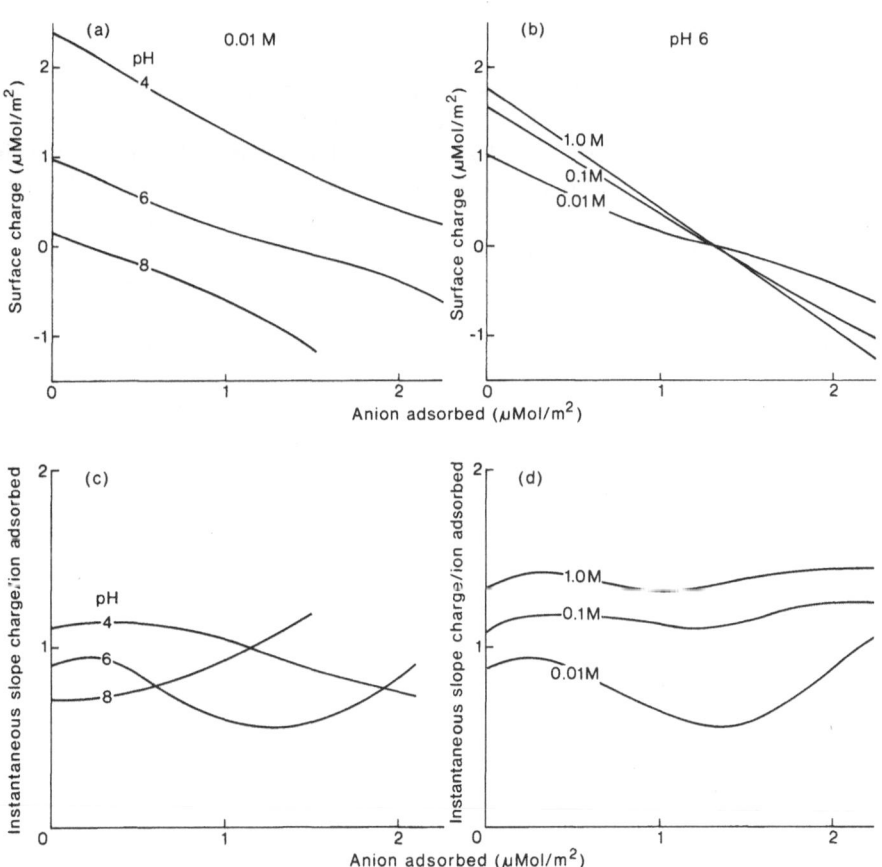

Fig. A5.5   Modelled effects of increasing amounts of adsorption of a divalent anion on the surface charge and
on the slope of the relation between charge and adsorption. The left hand parts show the effects at three pH
values; the right hand parts show the effects at three concentrations of a 1:1 electrolyte.

of minimum slope as they pass through the point of zero charge. These effects are most marked when the electrolyte concentration is low (Fig. A5.5b). These effects become clearer when the instantaneous slopes of these lines are plotted. These slopes reflect the effects on charge of each extra increment of adsorption. Fig. A5.5c shows that, at pH 6, the values pass through a minimum. Comparison with Fig. A5.5a shows that this is near the point of zero charge. This behaviour is analogous to that described in Fig. A5.4e and arises from the same cause — that is, differences in the way the charge is balanced for a positive surface, a neutral surface and a negative surface. At pH 4 only the approach to this minimum is observed because insufficient adsorption occurs to pass through the point of zero charge. At pH 8 only the climb from the minimum is observed because the surface is already negative at low levels of adsorption. These effects interact with electrolyte concentration in an analogous way to that in Fig. A5.4e and for an analogous reason.

Summary

This chapter outlines some of the effects of varying the parameters of the model and of varying some of the independent variables on the output of the model. Such exercises provide some familiarity with the model and a feel for how it works. However modelling should be a "do-it-yourself" activity. There is no better way to understand such models than to use them to ask questions like "What would happen if...?" This chapter should be regarded as a starting point for such activities rather than a complete answer.

References

Barrow, N.J., Bowden, J.W., Posner, A.M. and Quirk, J.P. 1980. Describing the effects of electrolyte on adsorption of phosphate by a variable charge surface. Australian Journal of Soil Research 18, 395-404.

Barrow, N.J. and Bowden, J.W. 1987. A comparison of models for describing the adsorption of anions on a variable charge mineral surface. Journal of Colloid and Interface Science (In Press).

Chapter A6

Fixed charge and variable charge components in soil

Before we could understand, and model, the reaction of nutrients and pollutants with variable-charge oxides, it was necessary to understand the charge on the oxides. When we come to the problem of soils we again face the problems of charge. In this case, we not only have variable-charge materials, but also fixed-charge materials. The questions which then arise are: is it possible to measure the charges on these different kinds of materials and is it necessary and profitable to do so? Before we can attempt to answer these questions we need to consider these two types of charged surface in more detail.

One of the characteristics of the clay minerals is that the silicon-oxygen sheets that comprise the faces of the crystals make a neat stopping point. This is unusual. For most crystals, the atoms on the external faces are unable to complete the pattern of the bulk of the crystal and they therefore have unsatisfied bonds. Metal atoms near the surface of crystals of metal oxides tend to satisfy their bonds by reacting with water molecules from the medium. In very simple terms, these water molecules may be thought of as either gaining or losing a proton — depending on the pH. This means that the surface acquires a charge and that the charge varies with pH. This mechanism does not occur for most of the surface of the clay minerals. Instead the charge derives from isomorphous substitutions within the crystal. Thus, the mechanisms of permanent-charge and variable-charge in soil are different. Unfortunately we tend to equate permanent-charge with clay minerals and variable-charge with metal oxides. It is more complex than this. Clay minerals always have edge sites as well as face sites. The edge sites also have unsatisfied bonds and so can have variable charge. On the other hand, oxides can have substitution of titanium for iron (Tessens and Zauyah 1982) giving rise to permanent positive charge. Further, if an oxide is permitted to react with, say, phosphate the charge on the oxide becomes more negative. That is, ions other than H^+ and OH^- can cause the charge to vary. Under some conditions, it can be difficult to remove some of the phosphate. In this case, the charge due to the phosphate could also be regarded as, at least fairly, permanent. Oxides in soil contain appreciable silicon and phosphorous (Chapter A2). These probably make the charge more negative and must also be regarded as fairly permanent. Thus the distinction between permanent charge and variable charge is not quite as sharp as it is sometimes portrayed.

A further complexity for soils is that charge on soil organic matter also varies with pH. The mechanism is somewhat different to that for oxides. In simple terms, the many different kinds of functional groups dissociate at different pH values. Some of them will be associated with a proton at low pH and be positively charged. Others will lose a proton at higher pH and be negatively charged.

The definition of surface charge

Some of the difficulties of defining and measuring surface charge arise from differing definitions of the surface. The problem is: if ions become arranged with some degree of regularity near the face of a crystal (Chapter A2), where is the boundary between the solid phase and the liquid phase — that is, where is the "surface"? One way of answering this question is to subject the particles to an electric potential in an electrophoresis apparatus. The charged particles will move in response to the potential. One can then argue that the entities that move are in fact the "particles" — that is, that the boundary between the particles and the liquid is the slip plane. If the particles are fairly close to spherical, the electric potential of this slip plane can be calculated. The problem is that, for oxides, the potentials so measured seem to be smaller than would be expected from the charge on the oxides. It is currently thought that some of the charge on the oxide is balanced by electrolyte ions (such as Na^+ and Cl^-) forming outer-sphere complexes with the oxide (Chapters A3 and A5). These ions then move with the oxide in response to the applied voltage. Because that which moves is defined as the particle, the surface is defined, in this case, as being outside these complexed ions. This is the definition used in much of the chemical literature. The difficulty for a soil scientist is that the charge on soil particles is usually measured by mixing samples of the soil with a solution of an electrolyte and then displacing the electrolyte and calculating the amount associated with the surface. It is implicit in this procedure that all the charge balanced by say Na^+ is surface charge. Thus, by this definition, the surface is *inside* the electrolyte ions. These two ways of thinking about the surface have led to a good deal of confusion about surface charge. The best way to clarify this is to introduce some symbols.

Let us define some subscripts for the various categories of charge near a surface. Thus:

v is the charge due to association and dissociation of protons from the surface;
p is the permanent charge;
a the charge due to adsorbed ions such as phosphate;
β the charge due to outer-sphere complexes;
d the charge in the diffuse layer.

The model of the distribution of ions near the surface implied by these definitions is now a common one (Chapter A3). The same model was also used by Sposito (1981) to clarify some of the definitions — though using different symbols.

Suppose we define the "surface" as in the first example above — that is, the surface includes the ions involved in outer-sphere complexes. Then the zero point of charge is defined by $\sigma_d = 0$. Sposito (1981) has used the symbol ZPC for this point. If, however, we define the "surface" as starting inside the ions forming outer sphere complexes then the point of zero charge occurs when $\sigma_\beta + \sigma_d = 0$. This point can be directly measured. The experiment involves mixing the soil with a solution of an electrolyte — say sodium chloride. Usually a fairly concentrated solution is used at first in order to remove other ions and then a more dilute solution. The centrifuged sample is weighed to measure the entrapped solution and then the ions are displaced with another electrolyte — say ammonium nitrate. The sodium and chloride ions held by the soil are then calculated and from this, the net charge. If the experiment is repeated at several pH values, a point may be reached at which the retained sodium and the retained chloride are equal. At this point the net charge is zero — that is, $\sigma_\beta + \sigma_d = 0$. Hence this point has been called the point of zero net charge (PZNC) (Sposito 1981). Note that this point is defined operationally; it is the point at which the measured net charge is zero.

Measuring the point at which $\sigma_d = 0$ is a little more complex and requires the introduction of another idea. If salt is added to a suspension of a variable-charge material, the pH changes. Alternately, if the pH were to be held constant, then either protons or hydroxide ions would have to be added to replace those taken up by the oxide — that is the charge would have changed. The direction of these effects depends on the charge on the surface. A middle point at which there is no effect may exist. This is the point of zero salt effect (PZSE). This point should also be defined operationally — simply as the point at which salt has no effect. However, under some conditions, it can be interpreted as the point of zero charge. The conditions are those for which σ_β is not affected by salt concentration. This could occur if none of the ions formed outer sphere complexes — and hence σ_β was zero. It also could occur if the electrolyte anion and the electrolyte cation both had equal tendency to form complexes. A change in the electrolyte concentration would produce a change in the amount of the complex but the net charge due to the complex would not change. It would remain zero. Under these conditions the point of zero salt effect is also the point of zero charge. If the electrolyte used to measure the net charge meets these criteria then the point of zero net charge would also be identical. To use the symbols, under these conditions, $\sigma_\beta = 0$, $\sigma_d = 0$, and PZSE = PZNC = PZC.

The effects of adsorbed ions

To date we have ignored the charge due to ions forming inner sphere complexes (σ_a). These include anions such as phosphate and fluoride, and cations such as copper and zinc. Suppose σ_a is appreciable and, for convenience, let us suppose an anion is involved so that σ_a is negative. The argument for a cation is similar but the signs are opposite. The presence of such a charge brings experimental problems — for example, it is difficult to measure the point of zero net charge without washing the soil with the salt solution. This could induce some desorption of the adsorbed anion and so a change in charge. Similarly the point of zero salt effect should be measured with the activity in solution of the adsorbed anion held constant. Supposing these difficulties are overcome, then, at the point of zero salt effect, $\sigma_v + \sigma_p + \sigma_a = 0$. If σ_a is negative, σ_v must be larger than it was before adsorption (more positive, or less negative) in order for the charges to sum to zero. This means that the proton concentration in solution must be higher. Thus the point of zero salt effect occurs at a lower pH. In summary, the presence of adsorbed anions, like phosphate, decreases the point of zero salt effect and the presence of adsorbed cations like zinc increases it. (The position is more complicated if the electrolyte solution contains ions which form inner sphere complexes — that is, it is not an indifferent electrolyte. This is treated adequately by Sposito (1981) and will not be considered further here.)

Separating permanent and variable charge

There is little controversy about the effects observed when σ_a is appreciable. However this is not the case when σ_p is appreciable. On one side of the controversy Uehara and Gillman (1980) have argued that, at the point of zero salt effect, σ_v is zero. Their argument is therefore that, if one measures the charge at this pH, then this charge must be equal to σ_p — the permanent charge. Thus their concept is that the surfaces present can be treated as consisting of two separate components — the variable charge components for which:

$$\sigma_v + \sigma_a + \sigma_\beta + \sigma_d = 0 \qquad\qquad\qquad A6.1$$

and the fixed charge components for which:

$$\sigma_p + \sigma_\beta + \sigma_d = 0 \qquad\qquad\qquad A6.2$$

This concept corresponds to the experimental system of Madrid *et al.* (1984). They mixed the iron oxide lepidocrocite with the clay mineral illite in various proportions. They then measured the net charge and the point of zero salt effect. The illite was negatively charged and so the net charge became more negative as the proportion of illite increased. Consequently the point of zero net charge occurred at lower pH values. However the point of zero salt effect was determined almost solely by the iron oxide and occurred at virtually the same pH value whether there was 100 per cent iron oxide or 10 per cent. Hence the charge present at that pH was due to the illite.

The controversial point is what happens if the variable charge and the fixed charge are not clearly separated so that, in the extreme case, we could write:

$$\sigma_p + \sigma_v + \sigma_a + \sigma_\beta + \sigma_d = 0 \qquad\qquad\qquad A6.3$$

In this case, a negative value for σ_p would have a similar effect to a negative value for σ_a: it would cause a lowering of the point of zero salt effect. This is quite contrary to the ideas of Gillman and Uehara (1980). This situation was approached experimentally by Hendershot and Lavkulich (1983). They precipitated iron oxide and aluminium oxide in the presence of clays and of soil materials. They showed that the oxides occurred as coatings on the minerals. The point of zero salt effect of the iron oxide alone was near pH 8.8. However in the presence of the other materials there was a wide range of values. They concluded that "there is no basis, either experimental or theoretical, for the assumption that the PZSE, as determined by potentiometric titration, is a measure of anything other than the properties of the system as a whole".

Thus, quite opposite effects were observed for two model systems both intended to mimic the behaviour of soil. The question then raised is which is the more realistic. Electron microscropy of soil often suggests close association between oxides and clay particles and one would expect that the edge sites of clays would be influenced by the permanent charge of the clay. It is therefore difficult to see how the permanent and the variable charge can be regarded as physically separate as required by Uehara and Gillman's model.

Is it important to distinguish permanent and variable charge?

The previous section suggests that it may be very difficult to distinguish between permanent and variable charge. Before too much effort is devoted either to disputing this conclusion or to agreeing with it and therefore trying to find better methods, it may be wise to ask whether it matters. There are three reasons for arguing that it doesn't matter. Firstly, from the academic point of view, the differences between the two kinds of charge are not as sharp as is sometimes portrayed. This seems to weaken arguments that it is important to measure them. Secondly, for the reaction with non-specifically sorbed nutrients such as potassium, the source of the charge is irrelevant. The important information is the total charge under realistic conditions of ionic strength and pH — and perhaps how that charge changes with pH. Thirdly, for the reaction of specifically sorbed nutrients such as phosphorus the problems of charge can be circumvented. This is considered in the next chapter.

References

Hendershot, W.A. and Lavkulich, L.M. 1983. Effect of sesquioxide coatings on surface charge of standard mineral and soil samples. Soil Science Society of America Journal 47, 1252-1260.

Madrid, L., Diaz, E. and Cabrera, F. 1984. Charge properties of mixtures of minerals with variable and constant surface charge. Journal of Soil Science 35, 373-380.

Sposito, G. 1981. The operational definition of the zero point of charge in soils. Soil Science Society of American Journal 44, 292-297.

Tessens, E. and Zauyah, S. 1982. Positive permanent charge in oxisols. Soil Science Society of American Journal 46, 1103-1106

Uehara, G. and Gillman, G.P. 1980. Charge characteristics of soils with variable and permanent charge minerals. I Theory Soil Science Society of America Journal 44, 250-252.

Chapter A7

The reaction of anions and cations with soil

One of our strongest urges is to classify things. As a result, an article on the chemistry of a particular nutrient often begins with a statement on the forms thought to be present in soil. For phosphate, for example, the forms may be stated as: calcium phosphates, adsorbed phosphates, occluded phosphates and organic phosphates. I want to avoid this approach. There are two reasons for this. One is that I have doubts about the importance of some of the postulated forms. For example, calcium phosphates may be rare indeed in many soils of the world (see later). The other is that the erection of sharp boundaries between postulated categories has a restricting effect on our thinking about the problem. It leads us to think in terms of these categories — for example to think in terms of "available" and "unavailable" nutrients. However, as we shall see, a better picture is that there is a continuum between these two ends of the spectrum. Thus, rather than start with a very simple model of the forms of the nutrients present in soil, I will start with a description of the observed behaviour. We will then be in a better position to consider a model to describe all of this behaviour.

An important characteristic of the approach is that the behaviour of several ions will be considered. Partly this is because of the general philosophy that, to be convincing, a model has to be comprehensive. And a comprehensive model must be able to describe all of the observed behaviour. It is also partly because comparison between ions can be a very powerful tool. For example, the fluoride ion is very useful for testing our ideas about multivalent ions like phosphate. Thus phosphate is thought to form a binuclear bond with oxides in soil. It has been suggested that the formation of more stable phosphate compounds involves the transition of mononuclear bonds to binuclear ones. This idea was first floated by Kafkafi *et al.* (1967) and is still current (Mengel 1985) even though it has been shown (Barrow and Shaw 1977a) that fluoride reacts in a closely analogous way to phosphate. Clearly fluoride could not form binuclear bonds. Thus, although phosphate may well form binuclear bonds, their formation may be fairly rapid and this is not the mechanism by which phosphate "ages". Despite the intention to describe the behaviour of several ions, many of the examples will be for phosphate. This is simply because phosphate has been of greater economic importance than the other ions and consequently more often investigated.

Some preliminary words on terminology are also needed. The word "specific" is used in a descriptive way. Thus, specific adsorption of zinc means that zinc has been adsorbed out of all proportion to its concentration relative to other cations. No mechanism is implied by the use of this term — but, of course, if the specificity is high one would suspect that a chemical bond had been formed. "Adsorbed" means that the ion is located on the surface of the solid phase. "Sorbed" is used in a much more general way. It merely means that the ion has been removed from solution. It may have been adsorbed, it may have penetrated the adsorbing surface to form a three-dimensional form, or it may have been precipitated. The term "adsorption isotherm" is avoided; the process described by this term is not merely adsorption and the use of "isotherm" wrongly implies that temperature is the only important variable apart from concentration.

The effects of concentration on sorption

One of the most-frequently studied aspects of the sorption of ions by soil is the effect of concentration in solution on the amount retained. There appear to be two main reasons for the interest in this subject. One is that the partition between the solution phase and the solid phase is one of the most important properties controlling the rate of movement in the soil. It therefore affects the rate of supply to plant roots, and both the rate of leaching from a profile and the amount of retention in a waste dump. Much effort has therefore been directed towards measuring the relation between concentration and sorption and in summarising the results of such measurements by equations. Most work has been done with phosphate and this was reviewed by Barrow (1978). Further contributions were made by Shayan and Davies (1978), Sibbeson (1981) and Mead (1981). Briefly, the situation is that the Langmuir equation usually describes sorption fairly well but with large and consistent deviations. One way of overcoming this is to use a sequence of Langmuir equations with different values for the binding constants and for the maximum sorption. On the pragmatic ground of efficient description this approach can seldom be justified. It is usually more efficient to use the Freundlich equation. Especially over limited ranges of concentration, this often describes sorption well and plots of log. sorption versus log. concentration approximate to straight lines. However, over wider ranges of concentration, these plots are more often gentle curves. Modifications of the Freundlich equation have been proposed to deal with this curvature (Shayan and Davies 1978; Sibbeson 1981). I have argued that the purpose of these equations is merely to summarize data and to permit interpolations (Barrow 1978). The choice between them should therefore be on the pragmatic grounds of simplicity and adequacy for the data to hand.

Although the above work was mostly concerned with phosphate, the generalizations also seem to apply to other anions. For metalic cations, such as zinc, the only modification required is that curvature of log.-log. plots may be more marked. At very low solution concentrations there appears to be a linear relation between solution concentration and metal retention (McLaren *et al.* 1983; Geritse and van Driel 1984). On a log.-log. plot this would give a slope of unity. With increasing concentration, the slope of the log.-log. plots decreases. This change in slope was described by Elrashidi and O'Connor (1982) by two linear segments each with a different slope. This seems to be another example of our proclivity to compartmentalize things rather than seek continuity.

The other reason for studying the relation between concentration and sorption is to try to understand the mechanisms involved. This reason has not always been separated from the previous reason and this has had two kinds of consequences. One is that supposedly mechanistic equations — such as the Langmuir — have sometimes been preferred on the ground that the parameters have physical meaning, even when the equation gives an inferior description of the data (and even though a Langmuir equation can never be appropriate for *ion* adsorption on a *charged* surface). The other is that conformity to a particular equation has been interpreted as evidence for a mechanism. For example, Holford *et al.* (1974) found that phosphate sorption could be described by a double Langmuir equation and concluded that phosphate was adsorbed on two types of surface with different bonding energies. Such a conclusion is untenable: the chemical mechanisms of sorption and the nature of the sorbing surfaces cannot be deduced from sorption curves alone (Sposito 1982). A mechanistic description of sorption requires that two criteria should be met. One is that our detailed knowledge of the reaction between ions and charged surfaces should be incorporated. The other criterion is that it should be recognised that the curve relating sorption and concentration is merely a two dimensional part of a multidimensional whole. The other dimensions include the effects of time, temperature and pH on sorption and also the effects on desorption. A convincing mechanistic description must include all of these effects. Before considering these further we will consider the relation between sorption and concentration for a calcareous soil.

When phosphate sorption is studied on a calcareous soil, the usual result is that, at first, the behaviour is not markedly different from that for a non-calcareous soil. Plots of sorption versus concentration form a typical curve. With increasing period of reaction, the concentration drops for all levels of addition but after a few hundred hours the drop in concentration becomes most marked when the level of addition of phosphate is high. With a long period of contact between soil and phosphate solution, treatments with a high level of phosphate all tend to converge to a common concentration of phosphate in solution of 1-2ppm P. A recent example of such an observation is that of Al-Khateeb (1986). This behaviour can best be explained by slow formation of a calcium phosphate. Freeman and Rowell (1981) showed that hemispherical, coral-like growths formed on portions of a calcium carbonate surface. Dicalcium phosphate was formed at first and slowly changed to octa-calcium phosphate. This is a clear case in which calcium phosphates were present. However we must be wary of extrapolating this kind of result to all soils — or even to all calcareous soils. The calcium phosphates form slowly, and moderately high phosphate concentrations are needed to initiate formation of dicalcium phosphate. A soil would have to have been fertilized with very heavy applications of phosphate before this process could be important. Even if calcium phosphates formed near the point of application of fertilizer, they would be expected to dissolve as diffusion moved phosphate to regions of lower concentration. There have been many attempts to identify calcium phosphates in soil by testing whether the concentrations in the soil solution corresponded to known solubility products. Usually the soil is found to be over- or under-saturated with respect to particular compounds. Further, if more phosphate is added, or the pH changed, the concentration seldom changes in a manner that suggests it is controlled by solubility products. One exception to these generalizations is the work of Weber and Mattingly (1970). They showed that when either lime or a calacareous soil were added to a soil of high phosphate status, the phosphate concentration in solution decreased until it appeared to be in equilibrium with octa-calcium phosphate. They concluded that octa-calcium phosphate appeared to impose an upper limit to phosphate concentration. Direct investigations of the soil solid phase have also provided little evidence for the widespread occurrence of calcium phosphates. Norrish and Rosser (1983) found that apatite grains could be observed in less-weathered Australian soils but, even in these soils, they were only a small fraction of the total phosphate.

With our modern knowledge of the reaction between phosphates and oxide surfaces, the meaning of the term "calcium phosphate" can be queried. Imagine that phosphate has reacted with an aluminous goethite and has formed one link to an iron atom and another to an aluminium atom in the goethite. The charge on this phosphate molecule will be balanced by ions in solution and it could well be balanced by a calcium ion. This phosphate molecule is therefore simultaneously an iron phosphate, an aluminium phosphate, and a calcium phosphate.

Effects of time on sorption

One of the characteristics of the reaction between soil and many nutrients is that it is rapid at first; it then becomes slower, but it continues for a very long time. Thus, a relation between sorption and concentration measured after a particular period represents only part of the interaction between soil and solution. Had a different period been used, a different relation would have been obtained. One therefore tends to wonder at the very large amount of effort devoted to fitting equations to data for just one period of reaction.

The continuing reaction with soil has been widely observed with phosphate but it also occurs with other ions — for example: molybdate (Barrow and Shaw (1975c); fluoride (Barrow and Shaw 1977a); sulfate (Barrow and Shaw 1977b; Singh 1984); arsenite (Elkhatib *et al.* 1984) zinc (Kuo and Mikkelsen 1979), and chromium, cadmium and mercury (Amacher *et al.* 1986). It has also been shown that incubating copper solutions with moist soil decreases the subsequent availability of copper to plants (Brennan *et al.* 1980, 1984). The continuing slow reaction is thus a very widespread phenomenon and may well occur for all specifically adsorbed ions.

The rate of any reaction can be measured by measuring either the rate of accumulation of products or the rate of disappearance of reactants. In the present case, the only practical measure is the rate of disappearance of the reactant. Thus the general experimental approach is to mix solutions containing the ion of interest with soil for extended periods and to measure the rate of decrease in solution concentration. There are two main variants of this general approach: in one, the volume of solution is large and the ratio of solution to soil is usually within the range of 5:1 to 100:1 ml g^{-1}; in the other, the volume of solution is merely sufficient to moisten the soil to its approximate field moisture content and the ratio of solution to soil is usually within the range of 0.1 to 0.5:1 ml g^{-1}. Though seemingly simple, both approaches pose some problems and both have advantages and disadvantages. These have been reviewed by Barrow (1983) and will be only briefly summarised here.

When a large solution: soil ratio is used, two experimental decisions must be taken. One is the background electrolyte to be used. Dilute solutions of calcium salts are widely used partly because they are similar to the composition of some soil solutions but mainly, I suspect, because soil and solution are easy to separate. Although 0.01 M calcium solutions are certainly much more concentrated than the soil solution of many soils, there is a good case for standardization so that results across a range of soils may be compared. There is already a large accumulation of data using such solutions and so there is a good case for regarding them as standard. Of course, for a particular study in which the aim was to make measurements relevant to plant uptake for the particular soil, there would also be a good argument for using a solution appropriate to that soil. The other experimental decision is the vigor of shaking. In many cases, vigorous end-over-end or reciprocating shakers are used. These are quite inappropriate for the slow reactions, and the consequent long periods of mixing. Especially for some soils, vigorous mixing can cause mutual abrasion of soil particles leading to misleading results for the rate of reaction. The mutual abrasion tends to be greatest when the solution:soil ratio is smallest thus leading to faster reaction. This effect can be avoided by using gentle methods of mixing (Barrow and Shaw 1979). We have found it effective to place bottles on their side on a roller that produces a few revolutions per minute.

An advantage of using large solution:soil ratios is that the solution can easily be separated for analysis. Indeed, where an appropriate electrode is available, sequential measurements can be made directly in the solution. However there are also disadvantages. The problem is that sorption is calculated from the measured change in solution concentration. Sorption and solution concentration are then treated as if they were independent variables even though one was, in fact, calculated from the other. The statistical problems that result are shown in detail in Chapter B7 together with a computer routine that circumvents the problem.

The use of a small solution:soil ratio has some advantages for measuring slow reactions. It is easier to store moist soil for long periods at constant temperature than it is to mix soil and solution for long periods. Further, the conditions under which the measurements are made are similar to those which would pertain if a plant was being grown. There are however some disadvantages. One possible disadvantage is that it may be necessary to use fairly high initial concentrations in solution in order to have a sufficient concentration for measurement at the end of the experiment. However the main problem is to measure the concentration in the soil solution at given times. Perhaps the simplest approach is to extract the soil solution by centrifuging but Helyar and Munns (1975) obtained variable results by this method. A better approach is to centrifuge the moist soil with a dense, immiscible liquid. The soil solution floats to the top and can be separated and analysed (Barrow 1982). Another approach is to extract the soil with a dilute salt solution such as 0.01M calcium chloride (Munns and Fox 1976). This results in a mixed experiment comprising an initial sorption phase followed by a desorption phase during the extraction. Especially when the sorption phase is short, this gives rise to complex kinetics. This problem can be avoided by using the "null-point" method. Instead of extracting the soil, it is mixed briefly with a series of solutions which contain a range of concentrations of the relevant ion. If the initial concentration is too low, the concentration will increase on mixing with the soil because of desorption from the soil. If the initial concentration is too high, the concentration will decrease because of sorption by the soil. The concentration at which no change would have occurred can then be interpolated — that is, the null-point. We have used this method extensively and many of the results quoted from our work were obtained using it.

Describing the rate of the reaction

A range of approaches has been used in attempts to describe the observed rates of reaction. A common approach is to assume, either explicitly or implicitly, that the reaction may be represented as $S + I \rightleftarrows SI$ (where S represents a site and I an ion). The simplest possible treatment of such a reaction is to assume that $S \gg I$, and that the back reaction can be ignored. This gives the hypothesis that the rate is proportional to the concentration of I —that is, a first-order reaction. This simple approach was used, for example, by Aringhieri and Pardini (1983) to describe the reaction of OH^- ions with soil. As might be expected, it only described the results over a very brief period. If these simplifying assumptions are not made, the reaction may be considered as a second-order forward reaction opposed by a first-order back reaction. At equilibrium, such a process gives rise to the Langmuir adsorption equation and hence the kinetics are often referred to as Langmuir kinetics. Aringhieri *et al.* (1985) attempted to apply such equations to the reaction of copper and of cadmium with an Italian soil. They found that they did not describe the rate and that the rate apparently decreased with time.

There are two reasons why these simple approaches do not work. The main one is that the slowest step in the reaction sequence is not the reaction of an ion with a surface site. This reaction is therefore not the rate limiting step and does not control the overall rate of the reaction. Before developing this argument further, another, less-important reason will be considered. It is that reaction between an ion and a surface site is not the simple adsorption pictured by the above equation. Both the site and the ion will be charged. These charges will affect the rate of the reaction. This is discussed in Chapter A4 and the equations are given in Chapter B3. However these equations only apply to the first few minutes of reaction of goethite with phosphate. Although they explain puzzling observations such as the effects of pH on the initial rate of the reaction (Barrow **et al.** 1981), they do not explain the slow reaction that followed the initial reaction.

Several approaches have been made to the problem of describing the slow, continuing reaction of ions with soil. A common approach has been to add a second reaction. For example, Lin *et al.* (1983) postulated that an initial adsorption reaction was followed by a reversible reaction in which "labile" phosphate was converted into "non-labile" phosphate. There is of course an incongruity here in that it is postulated that "non-lable" phosphate can be reconverted to "labile" phosphate. Novak and Petschauer (1979), on the other hand, used a model in which an initial adsorption reaction was followed by an irreversible surface reaction. One of the motives for this kind of approach is that the resulting equations can be incorporated into continuity equations and used to describe the movement of a pulse input of phosphate into a soil column. Thus Novak and Petschauer (1979) rejected other approaches because they could not be easily used for such problems. However the difficulty with approaches in which a further reaction is postulated is that they only describe the raction over short periods — in the case of Novak and Petschauer's (1979) experiments, for up to 24 hours. The values of the coefficients used by Novak and Petschauer were such that, after a few days, almost all of the phosphate would be in the irreversible surface phase. This seems unrealistic because the phosphate would then be unavailable to plants. Hence it does not seem that their relation can be extrapolated to practical periods. The temptation is then to use more-elaborate schemes in order to match the data. Probert and Larsen (1972) discuss the merits and demerits of using a sequence of first order reactions. They point out the dangers of this approach when the implied products cannot be identified and that the longer the experiments run, the more terms that are required. They therefore preferred a simple, two-constant equation that is similar to equation A7.2 (see later). The possibility that the results could be explained by a series of reactions was also considered by Barrow and Shaw (1975b). They showed that an exponential distribution of individual reaction rates could generate the observed rates. However, the equations they used made no provision for the return reaction and were therefore not comprehensive.

Many of the authors who observed that surface equations could not adequately describe the rate of reaction postulated that a diffusion step was involved. Sometimes this has been offered more-or-less as a speculation and not tested. Sometimes it has been tested — but inappropriately. Under certain restricted conditions, the amount of substance diffusing is proportional to the square root of time. One of the restrictions is that the surface into which the material is diffusing is a plane. For diffusion into soil particles this is clearly not so. However, if the distance of diffusion is small relative to the particle size, this is an acceptable approximation. The other restriction is that the source concentration must be constant. This has not always been appreciated. For example Tambe and Savant (1978) postulated that the rate limiting step was from the solution to the solid phase and hence the source concentration was the solution. However, in a batch experiment this must vary through time. In other cases the postulate is that the rate-limiting step is from the surface of the particle into the interior. Even if the solution concentration were constant, the surface concentration should not be expected to be constant — electrostatic effects consequent on the diffusion of charged particles would be expected to modify the surface electric potential and hence the surface concentration. Further consideration of diffusion models will be deferred until we deal with mechanistic models.

The Elovich equation has been advocated for describing the rate of sorption by soil (Chien and Clayton 1980; Hingston 1982). This equation was originally considered to be empirical. However Atkinson *et al.* (1971) showed that it could be derived by assuming a heterogenous distribution of activation energy on the adsorbing surface. To test this equation, sorption is plotted against log. time. This has sometimes been found to give a good description of sorption (Chien and Clayton 1980). There are two aspects of this finding that merit further discussion. One is that Chien and Clayton (1980) only tested the equation for one level of addition of phosphate and suggested that the intercept and slope of the plot could be used to characterise the soil. However had more than one level of phosphate been used, they would have observed that these values varied with level and

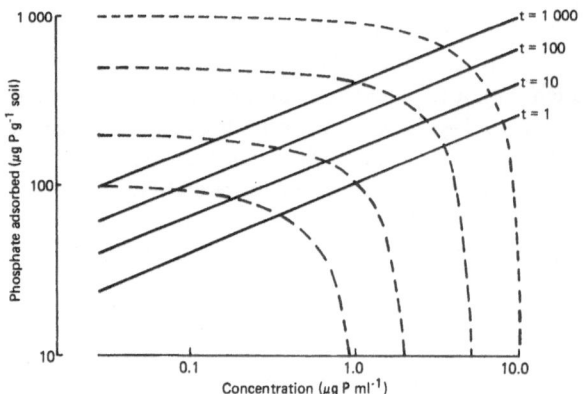

Fig. A7.1 Diagrammatic representation of the relationships when a soil that obeys equation A6.2 is mixed at a solution soil ratio of 100:1 with solutions with initial concentrations of 1,2,5 or 10 μg P ml^{-1}. The solid lines represent the equation with a = 100, b_1 = 0.4 and b_2 = 0.2. The broken lines connect points with the same initial concentrations of phosphate. The intersections of these lines with the solid lines indicate the predicted values for sorption and concentration at the indicated times (from Barrow 1983a).

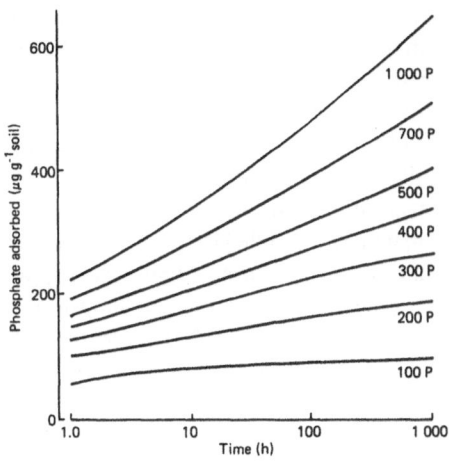

Fig. A7.2 Linear-log plot of predicted values for sorption obtained as in Fig. A7.1 for several levels of addition of phosphate. This method of plotting tests the Elovich equation.

were therefore not characteristic of a soil. The other aspect is that, under certain conditions, an Elovich plot can give a good description of part of the data even though a different equation applies to the whole of the data. This is illustrated in Fig.A7.1 and A7.2 in which it is assumed that equation A7.2 (see later) describes the data. Figure A7.2 shows that at low levels of addition, the Elovich plots bend to the right because all of the phosphate is soon sorbed. At high levels of phosphate the curves bend upwards because the limit due to the content of phosphate present becomes less important. At intermediate levels, however, the lines are nearly straight. Thus under these limited conditions an Elovich plot would appear to describe this data.

An alternative approach is to seek pragmatic functions that describe the rates over long periods. This approach does not require the initial setting up of a supposed mechanism. The philosophy is rather to explore and to describe what really happens. Armed with such measurements, and descriptions, we should then be in a better position to develop a hypothesis about the mechanism. An approach that has proved to be effective involves two assumptions. One is that *ad*sorption is rapid and that the consequent equilibrium between the adsorbed form and the concentration in solution can be adequately described by the Freundlich equation. The other is that the amount that remains in the *ad*sorbed form decreases with time (t) according to st^{-b_2} (where s is the sorbed form and thus includes the *ad*sorbed form and b_2 is a coefficient). Thus

$$ac^{b_1} = st^{-b_2} \qquad\qquad\qquad A7.1$$

and

$$s = ac^{b_1}t^{b_2} \qquad\qquad\qquad A7.2$$

This equation was introduced by Kuo and Lotse (1974) and was also used by Barrow and Shaw (1975a). In both cases the equation was derived from certain assumptions but it is probably better at this stage to regard it as a pragmatic equation.

Unfortunately, Kuo and Lotse (1974) modified the equation to:

$$s = ac_ot^{b_2} \qquad\qquad\qquad A7.3$$

where c_0 is the initial solution concentration. This relation was used to describe the effects of time on phosphate sorption by Schwertmann and Schieck (1980), on arsenite sorption by Elkhatib *et al.* (1984), and on sulfate sorption by Singh (1984). It describes a two-dimensional relation between sorption and time for each initial concentration. Thus it describes a slice across the three-dimensional relation. However it is not a slice at constant concentration because concentration inevitably changes as sorption proceeds. This is illustrated in Fig. A6.1. The figure shows that even though the diagram assumes a simple, three-dimensional equation, a separate equation would be needed for each different initial concentration if the simplified form of equation A7.3 is used. Thus the supposed simplification, in fact, leads to unnecessary complexity such as the calculation of rate coefficients at different times (Singh 1984) and the calculation of different coefficients for different initial concentrations (Schwertmann and Schieck 1980; Elkhatib *et al.* 1984).

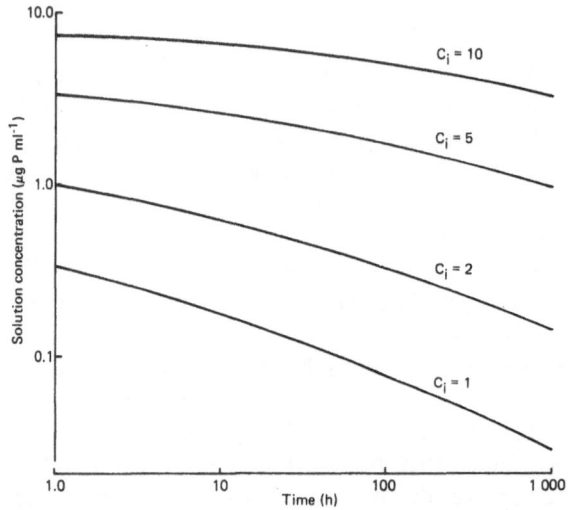

Fig. A7.3 Plot of expected values for concentration through time at four initial concentrations. Values were obtained as indicated in Fig. A7.1.

In contrast, equation A7.2 has only three coefficients. It has been found to be very effective in describing sorption over a wide range of conditions. For phosphate, molybdate and fluoride it closely described the effects of time over the equivalent of a 10,000 fold range of time (Barrow and Shaw 1975a,c, 1977a). (In these cases a range of temperatures was used to provide the equivalent of a range of times — see later). It describes phosphate sorption for a range of Australian soils (Barrow and Shaw 1975a) and a range of world soils (Barrow 1980a,b). It also describes sorption of sulfate (Barrow and Shaw 1977b). This wide applicability suggests that this simple, pragmatic equation somehow approximates to the mechanism. Of itself it tells us nothing of the mechanism — yet a mechanistic equation must be compatible with it. The wide applicability also suggests that the equation would be widely used in, for example, models of leaching. However differentiating equation A7.2 with respect to time gives:

$$\frac{\partial s}{\partial t} = ab_2 c^{b_1} t^{b_2-1} \qquad\qquad A7.4$$

Because b_2 is less than unity, this equation tells us that the rate of the reaction decreases with time — as is observed. It means that for any addition of phosphate or other nutrient, the current rate of change depends on the time that has elapsed since that addition was made. This behaviour is very difficult to incorporate in analytical models of leaching of solutes. To include it, one would have to use a less-sophisticated approach to modelling the movement of water — such as a layer model.

The unusual kinetics of the reaction pose some problems in discussing the rate of the reaction. To illustrate this, imagine that a soil was diluted with equal mass of an inert substance. At an equal concentration in solution, the mixture would retain only half as much sorbed material as the original soil. In terms of equations A7.2 and A7.4 the value of a would be only half as large as for the original soil. Hence we could say that the rate of sorption was only half as large. If analogous differences occurred between soils, plots of sorption (at constant concentration) against time could be made to coincide by adjusting the vertical scale — or by plotting s/a. This would compensate for the "amount" of sorbing material present. However, inspection of equations A7.2 and A7.4 shows that such plots would only coincide if the value of b_2 were the same. Variations in the value of b_2 reflect differences in behaviour through time. To illustrate this, consider the extreme values. A zero value for b_2 would indicate instantaneous reaction. A small value therefore indicates that the reaction is rapid at first and then becomes slow. A value of unity for b_2 would indicate a linear effect of time. Hence a large value of b_2 means that the reaction is relatively

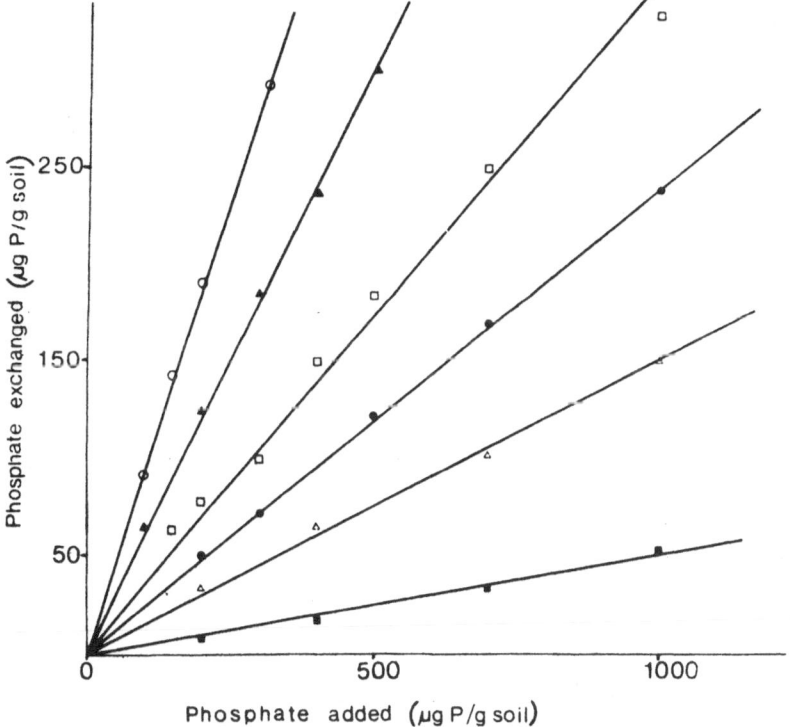

Fig. A7.4   Relation between the amount of phosphate added to a soil and the amount of phosphate that exchanged with ^{32}P. The separate lines represent differing periods of incubation with phosphate before measurement (from Barrow and Shaw 1975b).

slow at first but more persistent. Thus differences in the value of b_2 reflect different behaviour through time. When b_2 differs between soils it suggests that the soils differ in the 'kind' of sorbing material present.

The effect of level of application

It is convenient to regard a nutrient that may be sorbed as occurring in three categories: that in solution; that in the adsorbed form; and that in the third category which we will refer to as "firmly-held". The relation between the first two categories — solution and adsorbed — is non-linear as described earlier. However the relation between the second two categories — adsorbed and firmly-held — is linear. The proportion converted to the firmly-held form is independent of the level of application — or, in other words, the amount so converted is proportional to the amount added. This is implicit in equation A7.2 but is difficult to demonstrate using sorption data because of the non-linear relation between the first two categories. It can be better demonstrated by measuring the amount that is isotopically exchangeable after a given period as in Fig. A7.4.

Effects of temperature

Temperature may have two distinct effects on a chemical reaction: it may affect the rate of approach to equilibrium and it may affect the position of the equilibrium. Both these effects occur for the reaction of nutrients with soil. High temperatures of reaction increase the rate of reaction and give rise to decreased availability to plants, to decreased concentrations in solution, and to decreased subsequent desorption. Earlier references to these effects for phosphate were given by Barrow (1979a). More recent work by Chien et al. (1982) and by Sheppard and Racz (1984) have given similar results. In addition, analogous effects have been shown to occur for molybdate, fluoride and sulfate (Barrow and Shaw 1975c, 1977a, 1977b) and for zinc (Barrow 1986). Further, Brennan et al. (1984) has shown that high temperatures of incubation decrease the subsequent effectiveness of copper fertilizers. High temperatures during the desorption phase also increase the rate of the reverse or desorption reaction. Again earlier references to these effects were given by Barrow (1979) and similar results have been since described by Chien et al. (1982) and by Sheppard and Racz (1984).

In general, temperature affects the rate of a reaction because the reaction involves an intermediate, high-energy state. Only those molecules with sufficient energy can make the transition over this stage. The higher the temperature the higher the proportion of the molecules that have sufficient energy and so the faster the reaction. The effect of temperature is therefore conveniently described by the energy required to cross this barrier — the activation energy. For phosphate the activation energy for the forward (sorption) reaction appears to be similar to that for the backward (desorption) reaction (Barrow 1979a). Both were about 80kJ per mole. This suggests that the rate-limiting step is the same in both cases. It also suggests that the rate-limiting step is not diffusion in solution as this would require much less activation energy. Further, it indicates that temperature would have little effect on the position of an equilibrium reached by these rate-limiting steps. Yet if conditions are chosen such that the slow reaction has almost stopped and the temperature is then changed, high temperature is found to increase the concentration of adsorbing ions in solution. This effect must be on the position of the equilibrium of the *ad*sorption reaction. Earlier references on phosphate showing this effect are given by Barrow (1979a). A similar effect occurs for molybdate (Barrow and Shaw 1975c) and for zinc (Barrow (1986). These three separate effects of temperature are strong evidence for a model that postulates an initial rapid *ad*sorption reaction followed by a subsequent slow reaction.

If the rate of a reaction differs according to temperature then we should, perhaps, plot it against adjusted time, kt, rather than against time — where k is a factor to account for the different rates. Thus we might use k to equate a short period at a high temperature to a longer period at a low temperature. Equation A7.2 would then be written

$$s = ac^{b_1} (kt)^{b_2} \qquad\qquad A7.5$$

Note that the k term is enclosed within the bracket raised to the power b_2 because kt may be regarded as scaled time. In turn, k may be related to temperature using the Arrhenius equation

$$k = A \exp (-E/RT) \qquad\qquad A7.6$$

where E is the activation energy, R is the gas constant, T the absolute temperature and A is a constant. In most of the references above in which the value of E has been estimated it has been found to be about 80kJ per mole. At this value the rate of the reaction approximately triples for each 10° rise in temperature.

Repeated additions of a nutrient, and desorption

Suppose we incubate soil with a nutrient for an extended period. We might then choose to add more of the same nutrient and study the sorption of this repeated addition. Alternatively we might choose to attempt to remove some of the initially sorbed nutrient by, perhaps, decreasing the concentration in the solution — that is, we might choose to desorb some of it. These two operations are the opposite sides of the same coin and it is convenient to treat them together.

A simple modification of equation A7.1 has been proposed by Barrow and Shaw (1975a) to describe the effects of repeated additions. It is:

$$ac^{b_1} = s\, t^{-b_2} + s_1\, (t-t_1)^{-b_2} + s_2\, (t-t_2)^{-b_2} \qquad\qquad A7.7$$

where an amount s is sorbed at $t = 0$ and an extra amount s_1 is added at t_1 and s_2 at t_2 and where t is the total time. This equation assumes an additive effect with each addition of nutrient being treated as a separate pulse. Fig.A7.5 shows that it described fairly well the subsequent changes in concentration when repeated additions of phosphate were made to a soil. However, close inspection will show that the observed points all fall slightly above the lines predicted from equation A7.7. This effect arises because previous additions of phosphate affect the instantaneous slope of graphs of sorption versus concentration. This

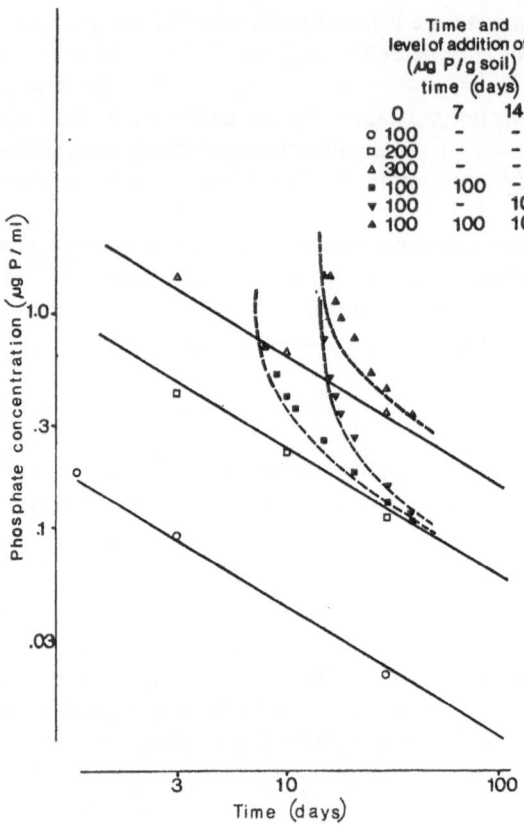

Time and
level of addition of P
(µg P/g soil)
time (days)

	0	7	14
o	100	−	−
□	200	−	−
▲	300	−	−
■	100	100	−
▼	100	−	100
▲	100	100	100

Fig. A7.5 Effect of repeated applications of phosphate on the subsequent
changes in concentration of phosphate
in solution. The solid lines indicate
equation A7.2 which was fitted to values
for a single addition. The values of the
parameters thus obtained were then
applied to equation A7.7 to predict the
concentration when repeated additions
were made (from Barrow and Shaw
1975a).

is illustrated in Fig.A7.5 which shows that the previous additions have a non-linear effect
on the slope of the sorption curves. The decreasing slope of the lines of Fig.A6.6 show that
this effect increases with increasing period of contact. It is therefore apparently caused by
the slow reaction that transfers some of the phosphate into a more-firmly held form. These
effects have mostly been studied with phosphate but similar effects occur with fluoride
(Barrow and Shaw 1977a) and so it seems feasible that they are general. Hence equation
A7.7 is too simple and should be modified to permit the value of a to vary according to the
amount of the original application that is now firmly-held. To illustrate this, let us consider
only one subsequent addition of nutrient and introduce a term to change the value of a
when the reaction of this addition is being considered:

$$c^{b_1} = st^{-b_2}/a + s_1 (t-t_1)^{-b_2}/k_1 a \qquad\qquad A7.8$$

where $k_1 < 1$. The introduction of the term k has the effect of increasing the value of c for
the subsequent additions.

Consider now the alternative of inducing desorption — that is a decrease in the amount
of nutrient instead of an increase. Rearranging equation A7.8 gives

$$s_1 = k_1 a c^{b_1} (t-t_1)^{b_2} - k_1 s \left(\frac{t-t_1}{t}\right)^{b_2} \qquad\qquad A7.9$$

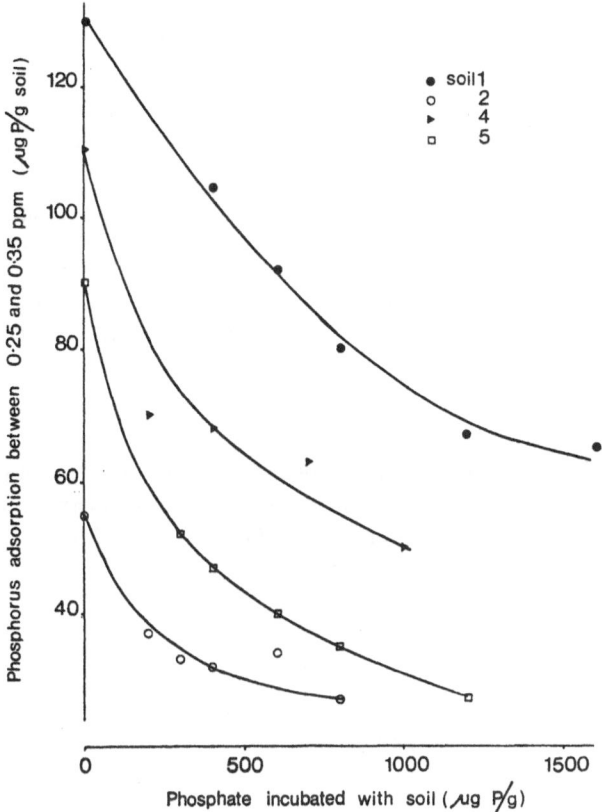

Fig. A7.6   Effect of incubating the indicated levels of phosphate with
soil for a year on the slope of graphs for subsequent sorption of
phosphate (from Barrow 1974).

Because desorption is the opposite of sorption, the amount desorbed (s_d) would be given by

$$s_d = k_1 s \left(\frac{t-t_1}{t}\right)^{b_2} - k_1 ac^{b_1}(t-t_1)^{b_2} \qquad\qquad \text{A7.10}$$

This equation can be used to generate typical sorption/desorption graphs (Fig.A7.7).
Further, this equation is similar to equations used by Barrow and Shaw (1975d, 1979b) to
describe desorption of phosphate and by Barrow and Shaw (1977a) to describe desorption
of fluoride. (The differences between the equations actually used and equation A7.10 arise
partly from approximations and partly because the initial sorption step was produced
using incubation conditions and a small solution:soil ratio whereas the subsequent step
used large solution:soil ratios and mechanical mixing. This produced slightly different
kinetics in the two parts of the experiment.)
    Thus equation A7.1 can be expanded to accommodate both the effects of further
additions of nutrient and to describe desorption. In each case, the subsequent addition (or

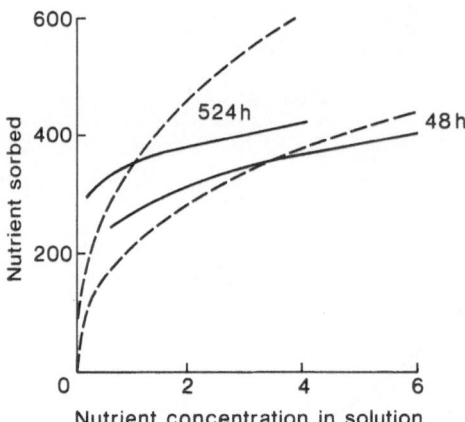

Fig. A7.7 Sorption and desorption graphs generated using equation A7.10. The broken lines indicate null-point concentrations when t = 48 and 524 hours; the solid lines indicate sorption and desorption for an initial level of 350 units of nutrient and for $t-t_1 = 24$. The arbitrary values for the parameters were; $a = 100$, $b_1 = 0.4$, $b_2 = 0.2$ and $k_1 = 0.7$ at 48 h and 0.5 at 524 h.

subtraction) is treated as a separate pulse which follows its own kinetic pathway. This treatment also shows that sorption and desorption are indeed symmetrical. This symmetry is shown when they are plotted on a common graph as in Fig. A7.8.

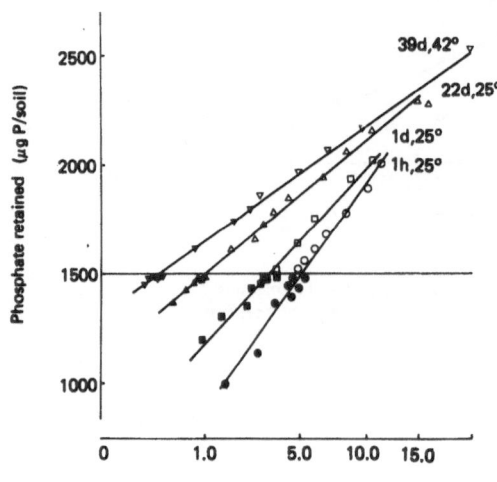

Fig. A7.8 Effect of incubating phosphate with soil at 1500 μg P/g for the indicated periods on the subsequent sorption (open symbols) and desorption (closed symbols) of phosphate. Data are plotted against concentration raised to the power b_1. This line arises equation A7.9 and the slopes reflect values of k_1a of equation A7.9 (from Barrow 1983).

This approach to desorption differs from that usually given in that it emphasises that the kinetics of nutrient reaction with soil are such that plots of desorption against concentration must follow a different path to plots of sorption against concentration — as in Fig. A7.8. Because these paths are different it is sometimes said that adsorption is irreversible. This is a contradiction in terms. By its nature, an *ad*sorption reaction cannot go to completion; there must be some of the adsorbate left in the solution phase and hence the back reaction must be appreciable. The term "irreversible", however, is usually used for a reaction that virtually goes to completion and there is virtually no back reaction — for example, the oxidation of hydrogen to give water. Thus the fact that the curves for

sorption and for desorption are different is clear evidence that the process involved is not merely adsorption.

To date we have merely assumed that desorption was induced — that is that the concentration in the solution phase was decreased and, as a result, some of the nutrient moved from the sorbed phase into the solution. In practice, such a decrease in the concentration in the solution could have arisen from uptake by a plant root. We turn now to consider how the concentration might be decreased in the laboratory and thus desorption studied. One widely-used technique, especially for phosphate, is to use exchange resins. Such studies provide much further evidence that desorption from soil is a slow process. For example the work of Amer *et al.* (1955), Moser *et al.* (1959) and Elrashidi *et al.* (1975) can all be used to show that release of phosphate is approximately proportional to a fractional power of time. However desorption to resins does not provide a good means of measuring the kinetics. This is because three distinct steps are involved — release from the soil, transport through the solution, and sorption by the resin — and each of these could be the rate-limiting step in different parts of the process. Transport through the solution and sorption by the resin are far from instantaneous. Thus the rate of sorption by the resin can be increased by decreasing the volume of solution or by increasing the vigour of mixing (Barrow and Shaw 1977c). Further, the rate is reduced if the resin is enclosed in mesh bags to make recovery easier. Thus, early in a run, when the release by the soil is rapid, the overall rate may be limited by the transport of ions through the solution to the surface of the resin. Later, when the release by the soil becomes slower this may become the limiting rate. Further, the nature of the exchange process is such that resins cannot decrease the solution concentration to zero (Bache and Ireland 1980). Indeed, as exchange proceeds, the equilibrium concentration will increase. Thus, through a run, the concentration in solution will vary in a complex way and, as is shown by equation A7.10, the rate of desorption is affected by the concentration in solution. It is this complexity that makes resin studies a poor way to study the kinetics of release by the soil.

An alternative, and simple, way to decrease the concentration in solution is to increase the volume of solution. The difficulty with this approach is that desorption is into a solution in which the concentration inevitably increases with time. This is not really avoided by repeatedly removing and replacing the solution (Kafkafi *et al.* 1967; Nagarajah *et al.* 1968) because, during each washing phase, the same criticism applies. In an attempt to overcome this problem Barrow and Shaw (1975d, 1979b) used a range of solution:soil ratios thus inducing a range of concentrations. By fitting functions like equation A7.10 they then hoped to interpolate the behaviour at constant concentration and to extrapolate to zero concentration. At any given solution:soil ratio, however, the concentration in solution must increase with time and equation A7.10 shows that this will have a feed-back affect and decrease the desorption. Thus, desorption observed at any given solution:soil ratio is always less than the maximum possible desorption. Further, the smaller the solution:soil ratio, the higher the subsequent concentration in solution, and hence the greater the reduction in desorption. These effects are illustrated for phosphate in Figure A7.9. This approach also allows us to explore the reversibility of sorption. Sorption is reversed when all of the sorbed phosphate is desorbed. Equation A7.10 shows that this will occur if the solution concentration is held at zero and when the term $k_1(t-t_1/t)$ approaches unity. That is, when the period of desorption is much larger than the period of prior

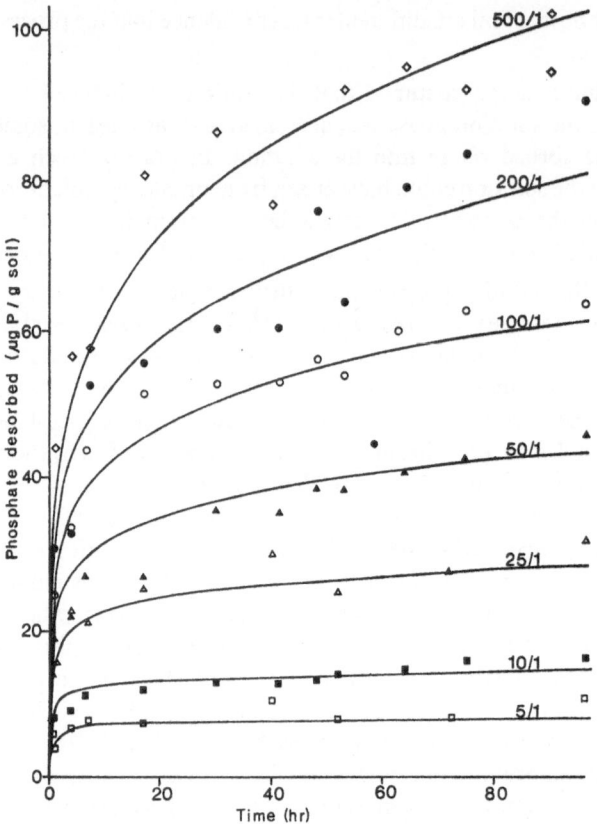

Fig. A7.9 Desorption of phosphate at a range of solution:soil ratios after
soil had been incubated with 1000 μg P/g soil for 18 days at 42°C. The
lines are fitted values from an equation like A7.10 (from Barrow and
Shaw 1975d).

sorption. At any given time, the reversibility of sorption is therefore given by the term
$k_1(t-t_1/t)$. However this equation is not appropriate to explore this further because of the
complex way that the parameter k_1 varies with the amount of sorbed phosphate. The
reversibility of sorption, however, can be explored using the regressions of Barrow and
Shaw (1979b) and extrapolating to zero concentration. Figure A7.10 shows that, after one
hour of prior contact, desorption is only 2/3 complete after 10h of desorption and only
approaches completion after 100h. After 22 days of prior contact, desorption is far from
complete after 100h. However there is no reason to suggest that, given sufficient time,
complete desorption would not be approached. Of course, this technique assumes that the
equation can be extrapolated to zero concentration and this is open to question.

Reversibility of sorption has also been measured in long-term field trials. In these cases,
desorption is induced by the decrease in concentration produced by plant roots. For
example, Leamer (1963) found that after 4 years of alfalfa and one of sorghum about 2/3
of the phosphate applied at 235kg/ha had been recovered. With subsequent crops,
recovery slowly increased and reached 80 per cent after nine years. At lower levels recovery

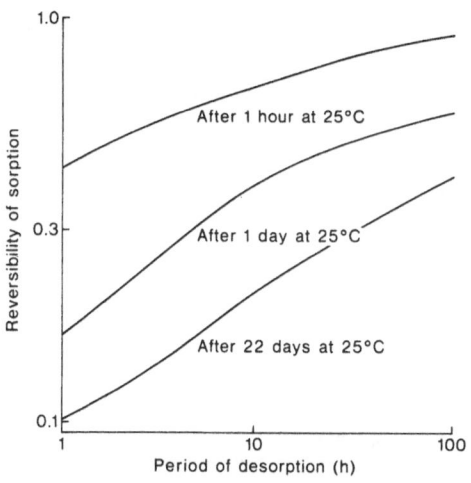

Fig.A7.10   Reversibility of phosphate sorption defined as the proportion of sorbed phosphate that would be desorbed if the solution concentration of phosphate could be held at zero for the indicated periods of desorption (calculated from the fitted regressions of Barrow 1979b).

was proportionately greater. In agricultural practice, fertilizer is re-applied well before such periods have elapsed because the rate of supply to plant roots has become too slow for good growth.

Desorption has also been widely studied in the presence of a range of reagents especially in the many soil tests applied to the diverse nutrients. This however is outside the scope of the present work.

Effects of pH and of electrolyte concentration

It is convenient to consider the effects of pH and of electrolyte concentration together because there is an interaction between them. This interaction is both the source of some of the controversy about the effects of pH and also an important clue in understanding the mechanism. It has been most studied for phosphate and consideration of it will be postponed until phosphate is considered. The other elements are considered first because the effects of pH seem to be more consistent. This really means that they are larger and so differences in the conditions of measurement do not have sufficient effect to change the direction of the effect of pH.

The effects of pH on sulfate and on molybdate sorption are much larger than those on phosphate. Between pH 4 and pH 6, the decrease in sulfate retention was about three times greater than that for phosphate while the decrease in molybdate retention was about twenty times greater (Barrow 1970). In contrast, boron retention increases with increasing pH (Hatcher et al. 1967). For most soils the maximum pH that can be reached is determined by the equilibrium between calcium carbonate and carbon dioxide. However the maximum pH for boron retention appears to be at a higher value than this. Sims and Bingham (1967, 1968a,b) showed that for soil constituents in sodium or potassium systems, maximum retention was near pH 9. A retention maximum near pH 9 was also reported by Keren and Mezuman (1981) and Keren et al. (1981).

Provided fluoride sorption is measured at equal concentration of total fluoride in solution, sorption is found to be greatest near pH 5.5 (Barrow and Ellis 1986a). Sorption is decreased at both low and at high pH. One special advantage in working with fluoride is

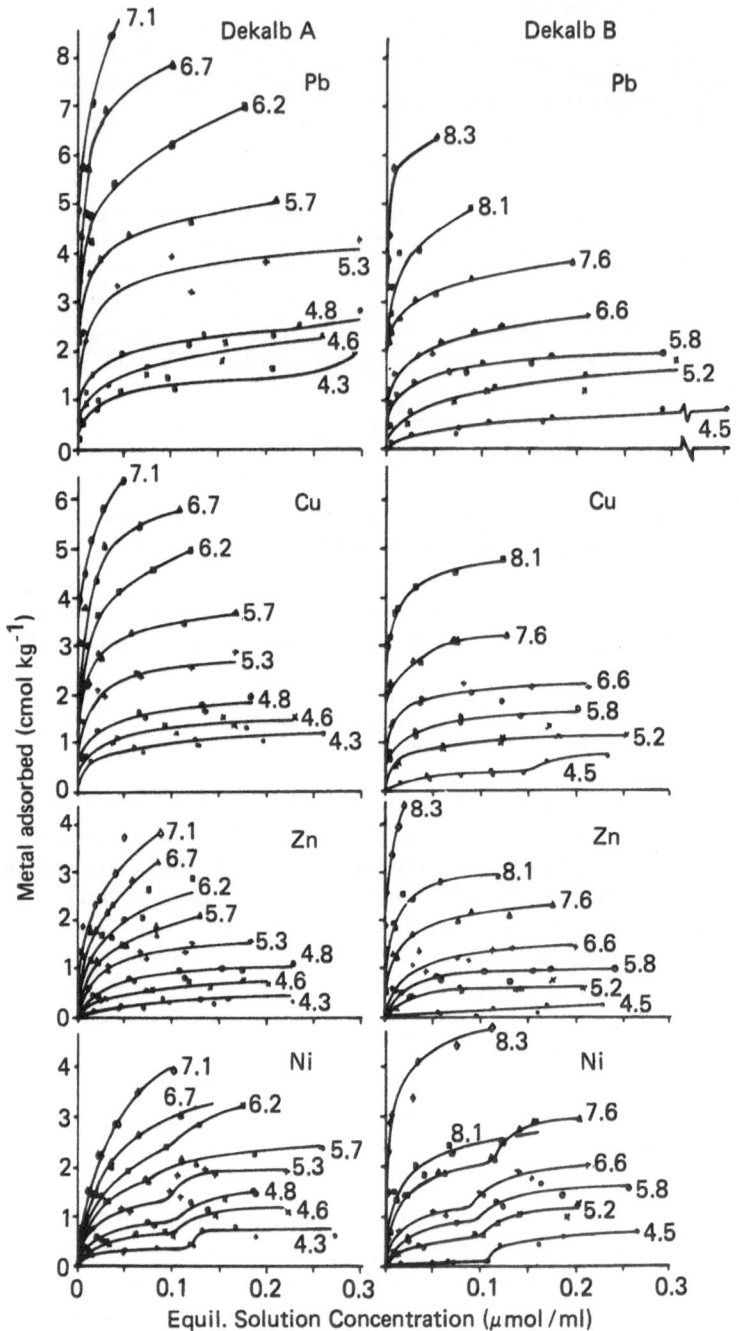

Fig. A7.11 Effect of pH on sorption of lead, copper, zinc and nickel by two soils (reproduced from Harter 1983 by permission of Soil Science Society of America, Inc).

that a specific ion electrode can be used to measure it. Such measurements show that at say pH 4, very little of the fluoride is present as F^-; most is present as complexes with aluminium. These complexes appear to greatly decrease the tendency of the fluoride to react with a soil and so decrease sorption. If sorption is related to free F^- in solution, rather than to total fluoride, sorption is found to decrease with increasing pH (Barrow and Ellis 1986a).

The possibility that the ion being considered has formed a complex with aluminium and/or iron ions in solution has often been overlooked. It is especially likely at low pH and is probably responsible for many reports that sorption decreases at low pH. For example Frost and Griffin (1977) reported that sorption of both arsenate and selenite by clays decreased as the pH was decreased below about 4. However, for both of these ions in the range of pH values normally expected in soils, sorption decreased with increasing pH.

For hydrolysable cations, all reports are constant in showing an increase in sorption with increasing pH. A comprehensive set of results for lead, copper, zinc and nickel was provided by Harter (1983) (Fig. A7.11).

In contrast to the fairly clear-cut results described above, the effect of pH on phosphate sorption has been controversial and seemingly contrasting results have been observed by different workers. Thus recent reviews have concluded: "Reports on the effect of liming on the sorption of phosphate are conflicting" (Probert 1980); "Considerable controversy exists in the literature regarding whether or not liming decreases P fixation" (Sanchez and Uehara 1980); and "Liming has been reported to increase, decrease, or not affect the phosphate that can be extracted from soils" (Hayes 1982). One overriding reason for this confusion may be suggested. It is that the effects of pH on phosphate sorption are smaller than the effects for other ions and so other factors become relatively more important. At least six factors have been proposed. These are discussed next.

The first factor to be considered is differences in mineralization of organic phosphorus. It is probable that increasing the pH of a soil increases the rate of mineralization. Further, the amounts of organic phosphate present will differ between soils and so these differences could lead to differences in the measured effects of pH (Haynes 1982).

The second suggested factor derives from the observation that increasing the pH decreases the aluminium concentration in solution and decreases the exchangeable aluminium. It has been argued that formation of hydroxyl-aluminium polymers occurs. This newly-formed material is postulated to be very effective in sorbing phosphate and differences between experiments are thought to occur because differences in handling — such as drying — give rise to different properties in the aluminium oxides (Haynes 1982). White (1983) has criticized some aspects of this suggestion but suported the general idea by suggesting that differences between soils in the effects of pH arise because soils differ in the amounts of exchangeable aluminium present. He suggested that, if a soil had more than 0.02 m mol g^{-1} of exchangeable aluminium, increasing the pH would increase phosphate sorption. There are two problems in evaluating this idea. One is that the exchangeable aluminium may, of itself, be effective in retaining phosphate. Thus Robarge and Corey (1979) showed that aluminium adosorbed on a cation exchange resin was effective in sorbing phosphate. The effectiveness varied with pH because of the changes in the aluminium species present and tended to decrease at high pH. Conversion of this material into a polymer would decrease the number of surface aluminium atoms and might

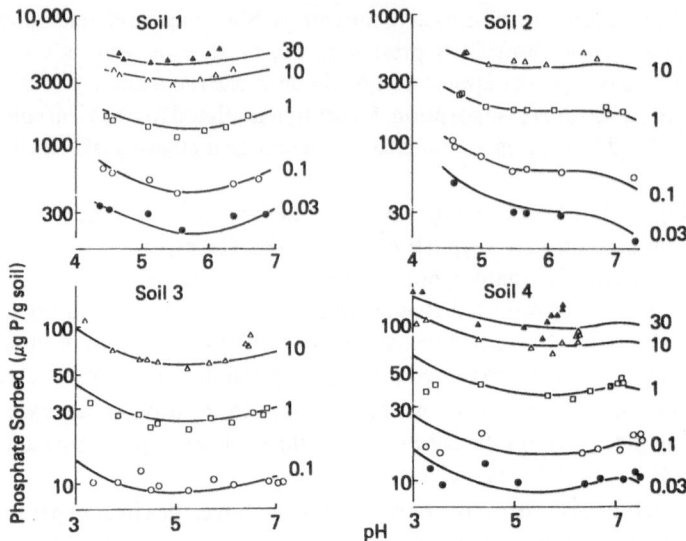

Fig. A7.12 Effect of changes in pH on the amount of phosphate sorbed at the indicated concentrations in solution by four virgin Australian soils. The lines drawn are from the model described in Chapter A8 (from Barrow 1984).

therefore be expected to *decrease* sorption. The other problem is to know whether the polymers formed could really be effective in sorbing much phosphate. For example, suppose we assume that the surface area of the polymer is the very large value of 10^3 m^2/g. Then 0.02 m mol g^{-1} would only have a surface area of 0.5 m^2. This is very small compared to the rest of the soil. This makes it at least questionable that formation of such polymers could be sufficient to oppose the effects of pH in making the variable charge surfaces more negative and so decreasing sorption — as suggested by Haynes and Swift (1985). Nevertheless this is a case in which a clear hypothesis has been put forward and one which can be tested. Suppose, for example, we were to take a soil with a large amount of exchangeable aluminium and raise its pH to say 7. If we then used the technique of Fordham and Norrish (1979) we could test whether there was any increased tendency for the reaction to be with aluminium rather than with iron and titanium atoms as observed by Fordham and Norrish.

The third factor is that the direction of the effect may depend on the pH range considered. Thus it is possible, in the same soil for the direction of the effect to differ at either end of the pH range (Fig. A7.12). Hence one researcher beginning at say pH 4 and raising the pH to say 5.5 would correctly conclude that raising the pH decreases phosphate sorption. Yet another researcher beginning at pH 5.5 and raising the pH to near 7 would (also correctly) conclude the opposite.

A fourth suggested factor is the presence of labile phosphate in the soil being examined. In this case two quite distinct effects of changing the pH could occur. One is an effect on the sorption of new phosphate — as would be observed in a soil with no labile phosphate. The other is the effect on the desorption of the labile phosphate in the soil. These need not be similar. For example it was found that the release of labile phosphate from a soil was most marked at low pH whereas the sorption of phosphate by a previously unfertilized sample of the same soil formed a "U" shaped curve against pH (Barrow 1984). Thus, raising the pH from a low value to a medium value decreased the release of labile phosphate but also decreased the sorption of new phosphate — that is, the effects were in opposite directions. If these differences are not understood, and adjusted for, it is possible for researchers studying two samples of the same soil, but with differing amounts of previously-applied fertilizer to reach opposing conclusions on the net effects of pH on phosphate sorption.

A fifth factor suggested as affecting the relation between pH and sorption is the effects of pH on the electrostatic potential in the plane of phosphate sorption. This relation will be dealt with in more detail when the mechanism of the reaction is considered. It suffices to note here that, for soils, we do not know *a priori* the relation between electric potential and pH. Hence if two soils had, for example, different mineral composition, the relation could differ. If the mechanism of the reaction is similar to that for pure variable-charge materials, such differences will change the relation between pH and sorption.

The sixth, and final, factor is the composition of the electrolyte in which sorption is measured. This effect can be substantial and failure to take account of it is certainly the cause of a good deal of confusion. For example the studies with oxides, for which increasing pH always decreases phosphate sorption, have been contrasted with the less-consistent effects with soils (White 1983). Yet studies with oxides have invariably used solutions of sodium or potassium salts as background electrolyte. These results are, in fact, consistent with those studies with soils which also used dilute sodium or potassium chlorides (Obihara and Russell 1972, Lopez-Hernandez and Burnham 1974; Parfitt 1977). A direct comparison on the same soil of the effects of sodium chloride with those of calcium chloride shows a much larger trend for sorption to decrease with increasing pH in the sodium chloride (Fig. A7.13). Further there is an interaction between the kind of electrolyte and the direction of the effect of pH. At high pH, sorption is higher in the calcium salt but at low pH it is lower. An interaction also occurs when the comparison is amongst sodium chloride solutions of differing concentration (Fig.A7.14). In this case, increasing the salt concentration increases sorption at high pH but decreases it at low pH. There is therefore a point at which the salt concentration has no effect on phosphate sorption — that is a point of zero salt effect on phosphate. This point of zero salt effect on

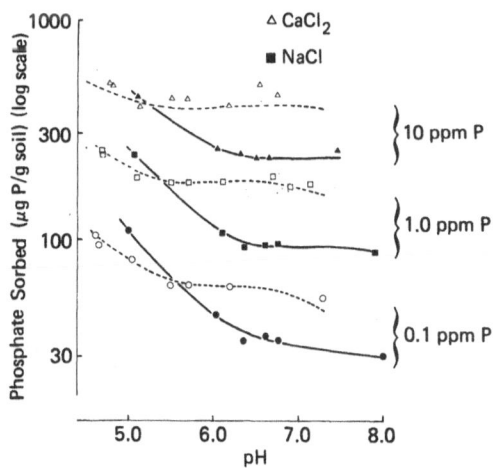

Fig. A7.13   Effect of background electrolyte on phosphate sorption at the indicated concentrations in solution (from Barrow 1984).

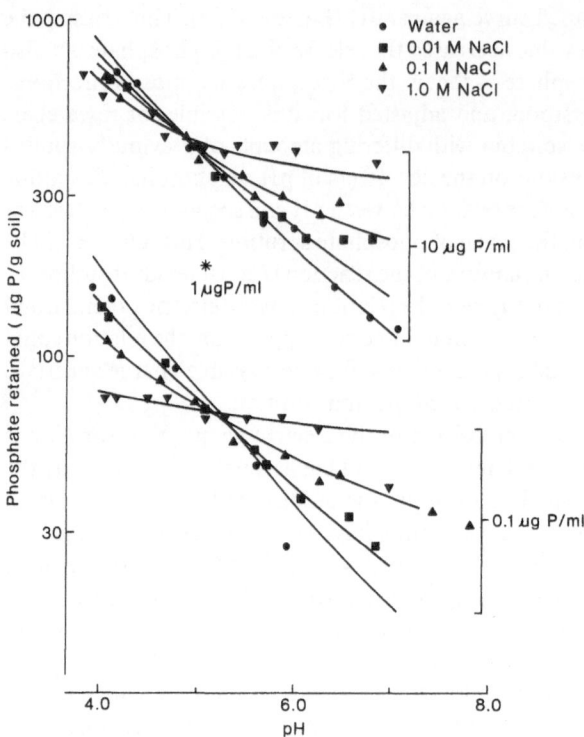

Fig. A7.14   Effect of concentration of sodium chloride on phosphate sorption at the indicated concentration in solution. The lines are drawn from the model described in Chapter A8 (from Barrow and Ellis 1986b).

phosphate decreases with increasing level of phosphate sorption (Fig. A7.15). Further it is at a higher pH than the point of zero salt effect on pH (Fig. A7.15).

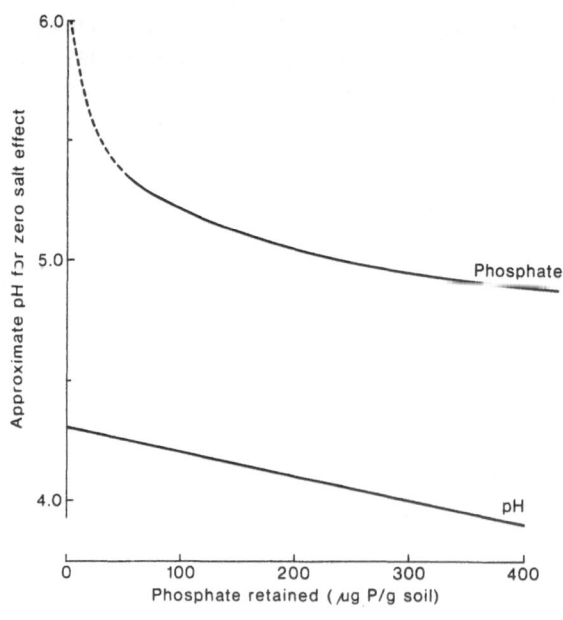

Fig. A7.15 Effect of level of addition of phosphate on the point of zero salt effect on phosphate sorption (as illustrated in Fig. A7.14) and on the point of zero salt effect on pH. The line for phosphate is drawn from the model described in Chapter A8 (from Barrow and Ellis 1986b).

These interactions between electrolyte and pH are important for three reasons. First they expand on the frequent observation that increasing the electrolyte concentration increases phosphate sorption by showing that the opposite effect occurs at low pH. The only satisfactory explanation is one that includes these effects. Secondly, because these effects must be due to differences in the way that the electric potential changes with pH in the different electrolytes, they support the general proposal above that such differences in potential could be important in explaining differences between soils. Thirdly, the existence of separate points of zero salt effect on pH and on phosphate sorption shows that phosphate is reacting with a subset of the total sites and that for that subset the mean electric potential is higher. This argument will be expanded in Chapter A8.

Summary

This chapter shows that reaction between nutrients and soils have some ten characteristics, all of which must be explained before an explanation can be considered to be comprehensive. The ten effects are summarised as follows.

1) There is a characteristic relation between amount sorbed and nutrient concentration in solution.
2) Reaction continues for a very long time at a rate that is proportional to a small, fractional power of time.
3) If this continuing reaction is thought of as slow conversion to a firmly-held form, the amount so changed is proportional to the amount added.
4) Increasing the temperature increases the rate of sorption.
5) Increasing the temperature increases the rate of desorption.
6) When the rates of sorption and desorption are slow, increasing the temperature changes the position of the equilibrium between solution and soil in such a way that the concentration in solution increases.
7) Repeated additions can be treated as separate pulses; they follow the same rules as single additions except that the instantaneous buffering capacity is decreased.
8) Desorption can be treated as a negative addition and the effects of desorption and of further sorption are symmetrical.
9) There are characteristic effects of pH for each nutrient.
10) The effects of pH interact with the nature and concentration of the background electrolyte.

References

Al-Khateeb, I.K., Raihan, M.J. and Asker, S.R. 1986. Phase equilibria and kinetics of orthophosphate in some Iraqi soils. Soil Science 141, 31-37.

Amacher, M.C., Kotuby-Amacher, J., Selim, H.M. and Ishandar, I.K. 1986. Retention and release of metals by soils — evaluation of several models. Geoderma 38, 131-154.

Amer, F., Bouldin, D.R., Black, C.A. and Duke, F.R. 1955. Characterization of soil phosphorus by anion exchange resin adsorption and ^{32}P equilibrium. Plant and Soil 6, 391-408.

Aringhieri, R., Carrai, P. and Petruzzelli, G. 1985. Kinetics of Cu^{2+} and Cd^{2+} adsorption by an Italian soil. Soil Science 139, 197-204.

Aringhieri, R. and Pardini, G. 1983. Interaction between OH⁻ ions and soil surfaces: a kinetic study. Canadian Journal of Soil Science 63, 741-748.

Atkinson, R.J., Posner, A.M. and Quirk, J.P. 1971. Kinetics of heterogeneous isotopic exchange reactions: derivation of an Elovich equation. Proceedings of the Royal Society A 324, 247-255.

Bache, B. W. and Ireland, C. 1980. Desorption of phosphate from soils using anion exchange resins. Journal of Soil Science 31, 297-306.

Barrow, N.J. 1970. Comparison of the adsorption of molybdate, sulfate and phosphate by soils. Soil Science 109, 282-288.

Barrow, N. J. 1978. The description of phosphate adsorption curves. Journal of Soil Science 29, 447-462.

Barrow, N. J. 1974. Effect of previous additions of phosphate on phosphate adsorption by soils. Soil Science 118, 82-89.

Barrow, N. J. 1979a. Three effects of temperature on the reactions between inorganic phosphate and soil. Journal of Soil Science 30, 271-279.

Barrow, N.J. 1979b. The description of desorption of phosphate from soil. Journal of Soil Science 30, 259-270.

Barrow, N. J. 1980a. Differences among some North American soils in the rate of reaction with phosphate. Journal of Environmental Quality 9, 644-648.

Barrow, N.J. 1980b. Differences amongst a wide-ranging collection of soils in the rate of reaction with phosphate. Australian Journal of Soil Research 18, 215-224.

Barrow, N. J. 1982. An evaluation of the immiscible displacement method for studying the reaction between soil and phosphate. Fertilizer Research 3, 423-483.

Barrow, N.J. 1983a. A discussion of the methods for measuring the rate of reaction between soil and phosphate. Fertilizer Research 4, 51-61.

Barrow, N.J. 1983b. On the reversibility of phosphate sorption by soils. Journal of Soil Science 34, 751-758.

Barrow, N.J. 1984. Modelling the effects of pH on phosphate sorption by soils. Journal of Soil Science 35, 283-297.

Barrow, N.J. 1986. Testing a mechanistic model. 2. The effects of time and temperature on the reaction of zinc with a soil. Journal of Soil Science 37, 277-287.

Barrow, N.J. and Ellis, A.S. 1986a. Testing a mechanistic model. 3. The effects of pH on fluoride retention by a soil. Journal of Soil Science 37, 287-295.

Barrow, N.J. and Ellis, A.S. 1986b. Testing a mechanistic model. 5. The points of zero salt effect for phosphate retention, for zinc retention and for acid/alkali titration of a soil. Journal of Soil Science 37, 303-310.

Barrow, N.J., Madrid, L. and Posner, A.M. 1981. A partial model for the rate of adsorption and desorption of phosphate by goethite. Journal of Soil Science 32, 399-407.

Barrow, N.J. and Shaw, T.C. 1975a. The slow reactions between soil and anions: 2 Effect of time and temperature on the decrease in phosphate concentration in the soil solution. Soil Science 119, 167-177.

Barrow, N.J. and Shaw, T.C. 1975b. The slow reactions between soil and anions: 3 The effects of time and temperature on the decrease in isotopically exchangeable phosphate. Soil Science 119, 190-197.

Barrow, N.J. and Shaw, T.C. 1975c. The slow reactions between soil and anions: 4 Effects of time and temperature of contact between soil and molybdate on the uptake of molybdenum by plants and on the molybdenum concentration in the soil solution. Soil Science 119, 301-310.

Barrow, N.J. and Shaw, T.C. 1975d. The slow reactions between soil and anions: 5 Effects of period of contact on the desorption of phosphate from soils. Soil Science 119, 311-320.

Barrow, N.J. and Shaw, T.C. 1977a. The slow reactions between soil and anions: 6 Effects of time and temperature of contact on fluoride. Soil Science 124, 265-278.

Barrow, N.J. and Shaw, T.C. 1977b. The slow reactions between soil and anions: 7 Effect of time and temperature of contact between an adsorbing soil and sulfate. Soil Science 124, 347-354.

Barrow, N.J. and Shaw, T.C. 1977c. Factors affecting the amount of phosphate extracted from soil by anion exchange resin. Geoderma 18, 309-323.

Barrow, N.J. and Shaw, T.C. 1979. Effects of soil:solution ratio and vigour of shaking on the rate of phosphate adsorption by soil. Journal of Soil Science 30, 67-76.

Brennan R.F., Gartrell, J.W. and Robson, A.D. 1980. Reaction of copper with soil affecting its availability to plants. I. Effect of soil type and time. Australian Journal of Soil Research 18, 447-459.

Brennan, R.F., Gartrell, J.W. and Robson, A.D. 1984. Reaction of copper with soil affecting its availability to plants. III. Effect of incubation temperature. Australian Journal of Soil Research 22, 165-172.

Chien, S.H. and Clayton, W.R. 1980. Applicability of the Elovich equation to the kinetics of phosphate release and sorption in soils. Soil Science Society of American Journal 40, 265-268.

Chien, S.H., Savant, N.K. and Mokwunye, U. 1982. Effect of temperature on phosphate sorption and desorption in two acid soils. Soil Science 133, 160-166.

Elkhatib, E.A., Bennett, O.L. and Wright, R.J. 1984. Kinetics of arsenite sorption by soils. Soil Science Society of American Journal 48, 758-762.

Elrashidi, M.A. and O'Connor, G.A. 1982. Influence of solution composition on sorption of zinc by soils. Soil Society of American Journal 46, 1153-1157.

Elrashidi, M.A. van Diest, A. and El Damaty, A.M. 1975. Phosphorus determination in highly calcareous soils by the use of anion exchange resin. Plant and Soil 42, 243-249.

Fordham, A.W. and Norrish, K. 1979. Arsenic-73 uptake by components of several acid soils and its implications for phosphate retention. Australian Journal of Soil Research 17, 283-306.

Freeman, J.S. and Rowell, D.L. 1981. The adsorption and precipitation of phosphate onto calcite. Journal of Soil Science, 32, 75-84.

Frost, R.R. and Griffin, R.W. 1977. Effect of pH on adsorption of arsenic and selenium from landfill leachate by clay minerals. Soil Science Society of America Proceedings 41, 53-57.

Gerritse, R.G. and van Driel, W. 1984. The relationship between adsorption of trace metals, organic matter and pH in temperate soils. Journal of Environmental Quality 13, 197-203.

Harter, R.D. 1983. Effect of soil pH on adsorption of lead, copper, zinc and nickel. Soil Science Society of American Journal 47, 47-51.

Hatcher, J.T., Bower, C.A. and Clark, M. 1967. Adsorption of boron by soils as influenced by hydroxyl aluminium and surface area. Soil Science 104, 422-426.

Haynes, R.J. 1982. Effects of liming on phosphate availability in acid soils. A critical review. Plant and Soil 68, 289-308.

Haynes, R.J. and Swift, R.S. 1985. Effect of liming and airdrying on the adsorption of phosphate by some acid soils. Journal of Soil Science 36, 513-521.

Helyar, K.R. and Munns, D.N. 1975. Phosphate fluxes in the soil-plant system: a computer simulation. Hilgardia 43, 103-130.

Hingston, F.J. 1982. A review of anion adsorption. In AndersonM.A. and RubinA.J. (Editors), Adsorption of inorganics at solid-liquid interfaces. Ann Arbor Science Publishers, Ann Arbor, Michigan.

Holford, I.G.R., Wedderburn, R.W.M. and Mattingly, G.E.C. 1974. A Langmuir two-surface equation as a model for phosphate adsorption by soils. Journal of Soil Science 25, 242-255.

Kafkafi, U., Posner, A.M. and Quirk, J.P. 1967. The desorption of phosphate from kaolinite. Soil Science Society of America Proceedings 31, 348-353.

Keren, R. Gast, R.G. and Bar-Yosef, B. 1981. pH dependent boron adsorption by montmorillonite hydroxyl-aluminium complexes. Soil Society of American Journal 45, 45-48.

Keren, R. and Mezuman, U. 1981. Boron adsorption by clay minerals using a phenomenological equation. Clays and Clay Mineralogy 29, 198-204.

Kuo, S. and Lotse, E.G. 1974. Kinetics of phosphate adsorption and desorption by lake sediments. Soil Science Society of American Proceedings 38, 50-54.

Kuo, S. and Mikkelsen, D. S. 1979. Zinc adsorption by two alkaline soils. Soil Science 128, 274-279.

Leamer, R.W. 1963. Residual effects of phosphorus fertilizer in an irrigated rotation in the southwest. Soil Science Society of America Proceedings 27, 65-68.

Lopez-Hernandez, D. and Burnham, C.D. 1974. The effect of pH on phosphate adsorption on soils. Journal of Soil Science 25, 207-216.

McLaren, R.G., Williams, J.H. and Swift, R.S. 1983. The adsorption of copper by soil samples from Scotland at low equilibrium solution concentrations. Geoderma 31, 97-100.

Mead, J. A. 1981. A comparison of the Langmvir, Freundlich and Temkin equations to describe phosphate adsorption properties of soils. Australian Journal of Soil Research 19, 333 - 342.

Mengel, K., 1985. Dynamics and availability of major nutrients in soils. Advances in Soil Science 2: 65-131.

Moser, U.S., Sutherland, W.H. and Black, C.A. 1959. Evaluation of laboratory indexes of adsorption of soil phosphorus by plants. Plant and Soil 10, 356-374.

Munns, D.N. and Fox, R.L. 1976. The slow reaction that continues after phosphate adsorption: kinetics and equilibrium in some tropical soils. Soil Science Society of America Proceedings 40, 46-51.

Nagarajah, S., Posner, A.M. and Quirk, J.P. 1968. Desorption of phosphate from kaolinite by citrate and bicarbonate. Soil Science Society of America Proceedings 32: 507-510.

Norrish, K. and Rosser, H. 1983. Mineral phosphate. In: CSIRO Div. of Soils (Editors), Soils: an Australian Viewpoint. CSIRO Melbourne/ Academic Press.

Novak, L.T. and Petschauer, F.J. 1979. Kinetics of the reaction between orthophosphate ions and Muskegon dune sand. Journal of Environmental Quality 8, 312-318.

Obihara, C.H. and Russell, E.W. 1972. Specific adsorption of silicate and phosphate by soils. Journal of Soil Science 23, 105-117.

Parfitt, R.L. 1977. Phosphate adsorption on an oxisol. Soil Science Society of America Journal 41, 1064-1067.

Probert, M.E. 1980. Growth responses to various calcium sources in a yellow earth soil with low calcium status. Australian Journal of Experimental Agriculture and Animal Husbandry 20, 240-246.

Robarge, W.P. and Corey, R.B. 1979. Adsorption of phosphate by hydroxy-aluminium species on a cation exchange resin. Soil Science Society of America Journal 43, 481-487.

Sanchez, P.A. and Uehara, G. 1980. Management considerations for acid soils with high phosphorus fixation capacity. In Khasawneh, F.E., Sample, E.C. and Kamprath, E.J. (Eds.). The role of phosphorus in agriculture. American Society of Agronomy. Madison, Wisconsin.

Schwertmann, U. and Schieck, E. 1980. Das Verhalten von Phosphat in eisenoxidreichen Kalkgleyen der Muenchener Schotterebene. Zeitschrift fur Pflanzenernaehrung und Bodenkunde 143, 391-401.

Shayan, A. and Davies, B.G. 1978. A universal dimensionless phosphate adsorption isotherm for soil. Soil Science Society of America Proceedings 42, 878-802.

Sheppard, S.C. and Racz, G.J. 1984. Effects of soil temperature on phosphorus extractability 1. Extractions and plant uptake of soil and fertilizer phosphorus. Canadian Journal of Soil Science 64, 241-254.

Sibbeson, E. 1981. Some new equations to describe phosphate sorption by soils. Journal of Soil Science 32, 67-74.

Sims, J.R. and Bingham, F.T. 1967. Retention of boron by layer silicates, sesquioxides and soil materials: I. Layer silicates. Soil Science Society of America Proceedings 31, 728-732.

Sims, J.R. and Bingham, F.T. 1968a. Retention of boron by layer silicates, sesquioxides and soil materials: II. Sesquioxides. Soil Science Society of America Proceedings 32, 364-369.

Sims, J.R. and Bingham, F.T. 1968b. Retention of boron by layer silicates, sesquioxides and soil materials: III. Iron and aluminium-coated layer silicates and soil materials. Soil Science Society of America Proceedings 32, 369-373.

Singh, B.R. 1984. Sulfate sorption by acid forest soils: kinetics and effects of temperature and moisture. Soil Science 138, 440-447.

Sposito, G. 1982. On the use of the Langmuir equation in the interpretation of "adsorption" phenomena II. The two-surface Langmuir equation. Soil Science Society of America Journal 46, 1147-1152.

Tambe, K.N. and Savant, N.K. 1978. Kinetics of sorption of orthophosphate and pyrophosphate by ammoniated tropical soils. Communications in Soil and Plant Analysis 9, 745-754.

Weber, M.D. and Mattingly, G.E.G. 1970. Inorganic soil phosphorus. II. Changes in monocalcium phosphate and lime potentials on mixing and liming soils. Journal of Soil Science 21, 121-126.

White, R.E. 1983. The enigma of pH-P solubility relationships in soil. 3rd International congress on phosphorus compounds proceedings. Institut Mondial du Phosphate, Casablanca, Morocco.

Chapter A8

Modelling the reaction of anions and cations with soil

The purpose of this chapter is to use the theories developed to describe the behaviour of variable charge materials in order to describe the behaviour of soils. The model presented will have two desirable properties: it will be comprehensive in that it will describe all of the ten characteristics of ion reaction with soils as listed in Chapter A7; and it will be based on knowledge of the details of the process as described in Chapters A3, A4 and A5. The model involves three assumptions. These will be described next.

The three assumptions of the model

1) *Ions react with variable charge surfaces*

The first assumption is, given the context, almost axiomatic. It is that ions in the solution react with the surface of soil particles that have a variable charge. For oxides, successful models for the reaction of ions with the oxide were developed from models of charge. These charge models involved using equations for the electric potential. It was the value for potential, so obtained, that was used in the equations that described adsorption. For soils, it does not seem to be a promising line of attack to work through charge models. Partly this is because there are problems in measuring variable charge in soil in the presence of fixed charge (Chapter A6); partly it is because soils contain a mixture of variable charge materials and the overall behaviour will be the sum of the separate parts; and partly it is because the variable charge material in soil is far from pure and these impurities will have effects on the charge and the potential as indicated in Chapter A4. Thus, charge models would have to be quite complex. Furthermore they are not really required. What is required are estimates of the values for electric potential and of the way that this changes with both changes in pH and changes in the amount of adsorption. That is, our soil model has to be consistent with the models for pure oxides but need not contain the same level of detail. Thus, for example, we may assume: that the electric potential decreases in a near-linear fashion with increases in pH; that the decrease is most marked for ions such as fluoride which would adsorb close to the s plane (Fig. A5.2); and that the decrease is most marked when the ionic strength of the background electrolyte is small (Fig. A5.4). Further we assume that adsorption of anions causes a linear decrease in the electric potential and adsorption of cations a linear increase. This assumption is taken from an approximation to the model for adsorption proposed by Posner and Barrow (1982). The change in potential arises from the change in charge shown in Fig. A5.5.

An ironical aspect of this approach is that emphasis is placed on the electric potential in the plane of ion adsorption yet accurate values cannot be allocated to it. There is no way to measure an appropriate value. Furthermore the values for potential occur in the equations for adsorption as a product with the term for the binding constant (Table A8.1). For soil, appropriate values for the binding constants for the various ions are also unknown. However what is really required is a value for the product. The approach adopted is to allocate a value to the binding constant. This value is somewhat arbitrary but is never-the-less kept in a range that is consistent with studies with oxides. The electric potential is thus treated as an adjustable parameter and values are allocated to it in order to match the data.

As was the case for the synthetic oxides, it is assumed that, of the several ionic forms that may be present in the solution, some are more strongly preferred than others. Indeed, in many cases, just one of the ions is sufficient to explain the observations. The strongest evidence for this part of the assumption is simply that it works — and thereby provides a consistent model to explain the behaviour of a diverse range of substances. In the case of phosphate another source of evidence is available. For phosphate, at pH values below 7, the monovalent ion predominates (Chapter A1) and there is a tendency to assume that this is the ion that is involved in reactions with the soil. However the model assumes that it is the divalent ion that dominates even though at say pH 4 it constitutes only a small fraction of the phosphate in solution. Somewhat more direct evidence for the importance of the divalent ion was obtained in a study of the displacement of molybdate from soil by phosphate solutions (Barrow 1974). Phosphate was most effective near pH 7 (Fig. A8.1). The increase in effectivness as the pH rises towards 7 can only be explained by an increasing concentration of the divalent ion. The decrease above 7 is explained by an

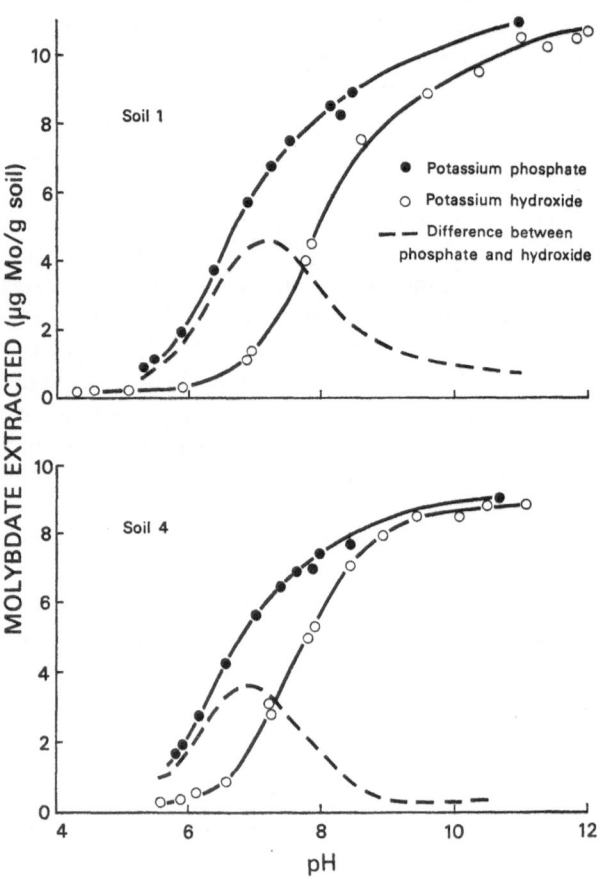

Fig. A8.1 Effect of pH on the molybdate extracted from two soils by solutions containing either 0.1 M phosphate or no phosphate. The broken lines indicate the effect of phosphate as obtained by difference. (From Barrow 1974)

increasingly negative charge on the surface coupled with little further increase in the concentration of the divalent ions. This observation is therefore taken as evidence that the divalent ion is important in reactions with soil.

2) *The surfaces are heterogenous*

The second assumption of the model is that the surfaces of the soil are not uniform. More specifically, there is a heterogeneity in their chemical properties. This assumption is needed to explain the characteristic shapes of graphs of solution concentration versus amount of sorption. It has been known for some time that the Freundlich adsorption equation can be generated if it is assumed that the surface is non-uniform and that there is an appropriate distribution of the values of the parameters of the adsorption equations (Zeldowitsh, 1935; Halsey and Taylor, 1947; Halsey, 1952). More recently, Sposito (1980) showed that one of the appropriate distributions is a normal distribution of the logs of the binding constants of a series of Langmuir equations. This is a convenient distribution to use in models. It can be readily expressed by an equation and, for modelling purposes, the total distribution can be subdivided into a series of slices. The properties of each slice are assumed to be represented by the characteristics of its midpoint, and the total behaviour is assumed to be that given by the sum of the slices. Further, the normal distribution can be specified by two parameters and these parameters have simple physical significance — the mean is the midpoint of the distribution and the standard deviation is, in simple terms, a measure of the spread of the values. A further advantage is that a normal distribution of the logs of the binding constants is similar to a normal distribution of the values for electric potential. This is because the terms for potential occur as exponentials in equations for adsorption. Thus the model allocates all of the heterogeneity to the potential term. This is primarily a device to simplify the model and should not be taken as a description of the real situation. Never-the-less there is evidence that, at least part of the heterogeneity is indeed associated with the electric potential. This will be presented later.

Figure A8.2 demonstrates how the assumed normal distribution can be used to describe sorption of phosphate. In this particular example, increasing the pH was found to decrease

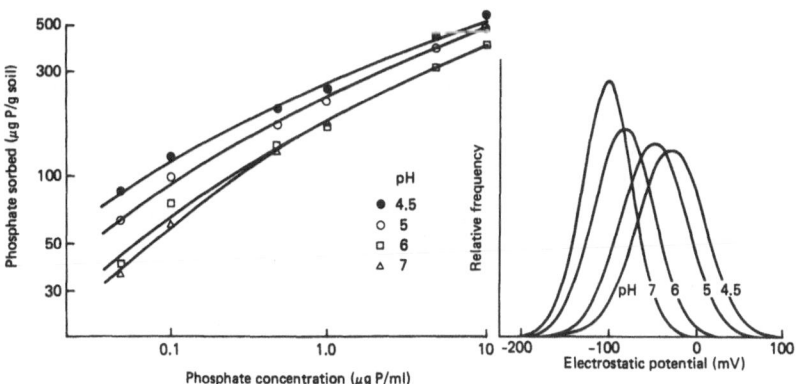

Fig. A8.2 Observed and modelled effects of concentration and of pH on phosphate sorption together with the distribution of electric potentials from which the lines were generated. (From Barrow 1983a)

the sorption but the magnitude of the effect differed at different concentrations. As a result, plots of sorption against concentration (on a log. scale) had a different slope at each pH. This behaviour was reproduced by assuming that the mean electric potential decreased with increasing pH but that the potential of the "outlying" slices decreased at a greater rate. This means that the standard deviation of the distribution decreased as pH increased. This is reflected by the narrower and taller distribution curves. Thus, a general rule is that a small value for the standard deviation generates a steep curve on a log.-log. scale. Note also that the model does not produce a straight line on a log.-log. scale but instead a gentle curve. This is probably more consistent with reality over a range of values for concentration.

An important part of this model for the effects of pH is that the concentration of the divalent phosphate ion increases with pH as indicated above.

3) *Adsorption is followed by diffusive penetration*

Because of the large amounts of evidence that there is a slow process that follows adsorption, a soil model has to provide a mechanism for this process. Although various reaction schemes have been proposed, there is only one that is consistent with all of the observations. It is that the initial adsorption of ions onto the surface generates diffusion into the bulk of the adsorbing particle. It perhaps needs to be emphasised that the postulate is that solid-state diffusion is occurring. Solid state diffusion has also been considered by Van Riemsdijk *et al.* (1984). Their model differs in two main respects from the present one. Firstly they do not consider electrical effects and hence the model cannot be comprehensive; it cannot account for effects of changing the background electrolyte or of changing the pH. Secondly their model postulates a precipitation-like process which only occurs when a certain concentration is exceeded. In contrast the present model sets no minimum concentration and this seems to be more consistent with observations.

Solid state diffusion is slow, and, because it is slow, it is the rate-limiting step. Although slow, the process is never-the-less more marked for soil than for oxides. It is argued that this occurs because the adsorbing material in soil is far from pure. In general terms, impurities in crystals are a source of strain and thus provide the irregularities that facilitate solid state diffusion. In more-specific terms, the silicon and phosphorus in soil goethites (Chapter A2) could provide the pathways for diffusion of anions.

As indicated in Chapter A4, equations do not seem to be available to describe the diffusion of ions from a charged surface into a crystal. The difficulty is that there is a gradient of electric potential into the crystal and this is a component of the gradient of electrochemical potential that induces diffusion. Diffusion equations should take this into account — but, at present do not. Faced with such uncertainty, the modeller's approach is to do the best he can with the tools to hand. This involves certain assumptions and approximations as indicated in Barrow (1983a) and in Chapter A4. These assumptions give rise to the equations given in Chapter A4. For convenience they are repeated here as A8.1 and A8.2:

$$M = (2/\sqrt{\pi})\, C\sqrt{(\tilde{D}f\, t)} \qquad\qquad A8.1$$

where M is the amount of substance diffusing into a crystal during the period t when the

84

surface concentration is C , and $\tilde{D}$ is a diffusion coefficient. The term f is known as the thermodynamic factor. It is derived from a function of the ratio of surface activity to surface concentration.

As indicated in Chapter A4, Equation A8.1 describes diffusion into a plane. That is, it is implied that the rate of diffusion is so slow that we can ignore the "radial convergence" of diffusion lines. Another aspect is that the term C refers to surface concentration measured in moles per area rather than moles per volume. This means that the dimensions of $\tilde{D}$ are time^{-1}.

Equation A8.1 can only hold while C is constant. An important further assumption of the model is that the diffusion of ions into the surface changes the electric potential of the surface. Consequently, the surface concentration C will change even though the solution concentration be kept constant. This effect of ion penetration on surface potential is modelled through a simple feed-back term (Equation (6) of Table A8.1). That is, over the range of values for penetration being considered, it is assumed that the effect on the potential in the plane of adsorption is linear. The process is then followed using a series of small steps and it is assumed that during each step C remains constant. To illustrate the required modification of Equation A8.1, let the initial surface concentration be C_o, and let this change at t_1 to C_1. Then at a subseqent time t, a linear superposition of equations like A8.1 gives:

$$M = 2/\sqrt{\pi} \; (C_o \sqrt{(\tilde{D} \; f \; t)} + (C_1 - C_o) \sqrt{(\tilde{D} \; f \; (t-t_1))}) \qquad\qquad A8.2$$

The change in concentration at time t_1 could have arisen from the change in surface electric potential or it could have arisen from a change in the solution concentration induced in order to produce either desorption or further sorption. Equation A8.2 appears in its more general form as Equation(6) in Table A8.1.

The formal equations of the model

The equations are gathered in Table 8.1 Section A describes the heterogeneity of the surface. It refers to components of the surface for which the subscript "j" is subsequently used. Section B presents the equations that describe adsortion on each of these components, either at equilibrium or during the approach to equilibrium of the adsorption step. Section C describes the diffusive penetration, and Section D the feed-back effects. The equations for the model are completed by Section E which presents those effects of temperature which are additional to those of Section B. Equation (7) shows that the diffusion coefficient is related to temperature through the Arrhenius equation and Equation (8) shows that the binding constant is also affected by temperature. There may be further effects of temperature on the ionic composition of the solution if, for example, the heat of reaction for dissociation is appreciable.

The model also requires a means of describing the changes in the electric potential in the plane of adsorption caused by changes in the pH. Pragmatic equations could be used for this but, in the applications of the model to date, smooth hand-drawn curves have been used. The curves differ for different ions and also differ for different ionic strengths of the background electrolyte.

Table A8.1 Tabulation of the equations that describe the model

A *Heterogeneity of the surface*

$$P_j = 1/(\sigma/\sqrt{2\pi}) \exp[-0.5((\psi_{a0j} - \bar{\psi}_{a0})/\sigma)^2] \tag{1}$$

P_j is the probability that a particular particle will have an initial potential of ψ_{a0j},
$\bar{\psi}_{a0}$ is the mean value of the initial potential,
σ is the standard deviation of the initial potential.

B *Adsorption on each component of the surface*

(i) at equilibrium

$$\theta_j = \frac{K_i \alpha \gamma c \exp(-z_i F \psi_{aj}/RT)}{1 + K_i \alpha \gamma c \exp(-z_i F \psi_{aj}/RT)} \tag{2}$$

θ_j is the proportion of the jth component occupied by the adsorbed ion
K_i is a binding constant for the ion i,
z_i is the valency of the adsorbing ion,
ψ_{aj} (mV) is the electrostatic potential of the jth component,
α is the fraction of the adsorbate present as the ion i,
γ is the activity coefficient,
c is the total concentration of the adsorbate,
F, R, and T have their usual meaning.

(ii) rate of adsorption

$$\theta_{jt} = \frac{K_1^* c(1 - \theta_j) - k_2^* \theta_j}{k_1^* c + k_2^*} [1 - \exp(-t(k_1^* c + k_2^*))] \tag{3}$$

θ_{jt} is the increment in θ_j over a time interval t, and

$$k_1^* = k_1 \alpha \gamma \exp(\bar{a} F \psi_{aj}/RT) \tag{3a}$$

$$k_2^* = k_2 \exp(-\bar{a} F \psi_{aj}/RT) \tag{3b}$$

k_1 and k_2 are rate coefficients and $\bar{a}$ and $\vec{a}$ are transfer coefficients.

C *Diffusive penetration*

$$M_j = 2/\sqrt{\pi} [C_{0j}\sqrt{(\tilde{D}ft)} + \overset{k}{\Sigma}(C_{kj} - C_{kj-1})\sqrt{(\tilde{D}f_k(t - t_k))}] \tag{4}$$

M_j is the amount of material transferred to the interior of the jth component expressed on an area basis,
C_{0j} is the surface concentration of the adsorbed ion at time t,
C_{kj} the value to which C_{0j} changes at time t_k,
$\tilde{D}$ is a coefficient related to the diffusion coefficient via the thickness of the adsorbed layer (the dimension of $\tilde{D}$ is time^{-1}),
f is the thermodynamic factor.
(These terms are defined in more detail in Barrow (1983a)).

D *Feedback effects on potential*

(i) For a single period of measurement

$$\psi_{aj} = \psi_{a0j} - m_1 \theta_j \tag{5}$$

ψ_{aj} is the potential of the jth component after reaction,
m_1 is a parameter

(ii) For measurement through time

$$\psi_{aj} = \psi_{a0j} - m_1 \theta_j - m_2 M_j/N_{mj} \tag{6}$$

N_{mj} is the maximum adsorption on the component j
m_2 is a parameter

E *Effects of temperature*

$$\tilde{D} = A \exp(-E/RT) \tag{7}$$

E is an activation energy
A is a parameter

$$K_1 = \exp(B/RT) \tag{8}$$

B represents potentials in specified standard states plus an interaction term (Bowden *et al.*, 1977)

Modelling the characteristics of sorption

Chapter A7 summarised ten observations as characteristic of sorption. In this section we will consider the ability of the model to match these observations.

1) *The relation between sorption and concentration*

Figure A8.2 shows that the model reproduces the characteristic gentle curve when the log. of phosphate sorption is plotted against log. concentration. The slightly curved nature of this plot is also shown for zinc in Fig A8.3. The model has also successfully reproduced the

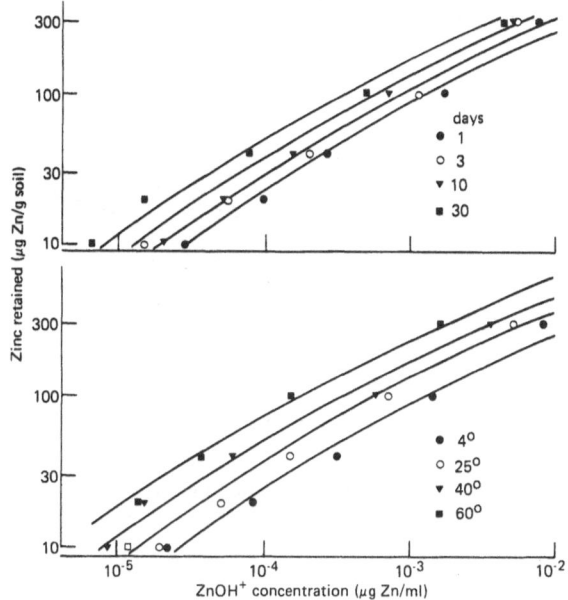

Fig. A8.3 Effect of period and of temperature of incubation on the relationship between zinc sorption and the concentration of ZnOH⁺ ions in solution. The upper figure indicates the effect of period at 25°C; the lower figure the effect of temperature after 10 days. The lines were fitted from the model. (From Barrow 1986b)

relationship between sorption and concentration for fluoride and for molybdate sorption (Barrow 1986a).

2) *The effects of time on the reaction*

The observation to be matched is that sorption, at constant concentration, is proportional to a fractional power of time. The model postulates that sorption is proportional to the square root of time but that the rate is decreased below this value by the feed-back effects of sorption on electric potential. That is, the feed-back decreases the fractional power below the value of 0.5 given by a square root term. Figures A8.3 and A8.4 show that the model closely describes the effects of time on zinc and on phosphate sorption. The effects of a greater range of times are shown in Fig. A8.5. In this case the simple description that the

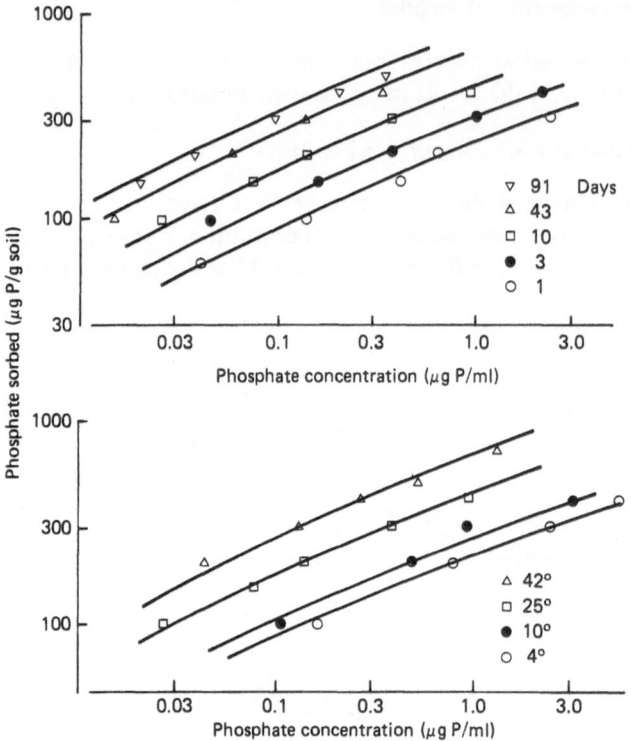

Fig. A8.4   Effect of period and of temperature of incubation on the relationship between phosphate sorption and phosphate concentration. The upper figure indicates the effect of period at 25°C; the lower figure indicates the effect of temperature after 10 days. The lines were fitted from the model. (From Barrow 1983a)

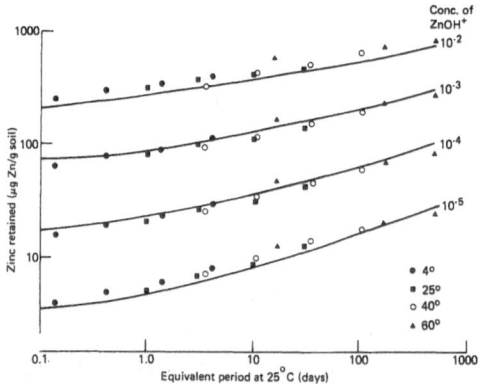

Fig. A8.5 Changes in observed and modelled sorption of zinc through time at the indicated concentrations of $ZnOH^+$. Periods spent at temperatures other than 25°C have been adjusted to periods at 25°C using Equation(7) of Table A8.1. (From Barrow 1986b)

effects are proportional to a fractional power of time is seen to be an approximation in that the log.-log. plots for the various concentrations are not parallel — as they would be if equation A7.2 were followed. However the model gave a close description of the observations. An important component of the model in reproducing these effects for zinc was a large value for the feed-back term (Barrow 1986b).

3) *The amount converted to a "firmly-held" form is proportional to the amount sorbed.*

For the purpose of illustrating this point it is neccessary to assume that the amount that has penetrated the adsorbing surface is an indication of the amount that is "firmly-held". Equations A8.1 and A8.2 show that the amount penetrated is proportional to the surface concentration.

Although it is sometimes convenient to describe a fraction of the sorbed material as being firmly-held, this is, in fact, too simple a description. In reality there is a range of "firmness". This is also reproduced in the model in that the range of firmness corresponds to depth of penetration.

4) *High temperatures accelerate sorption*

Figures A8.3 and A8.4 show that the model reproduced the effects of temperature on the rate of zinc sorption and on the rate of phosphate sorption. The model also reproduces the effects of temperature on the rate of fluoride and molybdate sorption (Barrow 1986a). In all cases, these effects were measured by incubating soil plus added chemical at different temperatures but measuring the solution concentration at a common temperature. This means that the effects of temperature on the equilibrium between the solution and the adsorbed form (Section B of Table A8.1) were not relevant. The only effect of temperature that was relevant was that due to Equation (7) of Table A8.1 — that is, the effect described by the activation energy for diffusion. The numerical value for this term was similar to that estimated from the pragmatic Equation A7.2 — about 80kJ per mole. This fairly-high value for the activation energy reflects the fact that diffusion in the solid state involves the movement of an ion from a low energy position, over an energy barrier, to a new low energy position. The activation energy is a measure of the height of that energy barrier.

Equation (7) of Table A8.1 may be used to give the relative rates at different temperatures and hence to calculate equivalent periods at a common temperature. This was the technique used in Fig. A8.5 in order to illustrate the process through a large range of "equivalent time". That figure shows that the measurements made at near-by temperatures overlapped — thus providing further evidence that an appropriate description of the effects of temperature was obtained.

5) *High temperatures accelerate desorption*

During the first few minutes of a desorption phase, desorption of adsorbed ions occurs. After this brief initial phase, further desorption depends on the return diffusion of some of the penetrated ions to the surface — that is, diffusion is again the rate-limiting step. The effects of temperature on the rate of this step are therefore the same as on the rate of the sorption step.

6) *High temperatures decrease adsorption*

The above two sections refer to the effects on the rate of sorption and thus (in the model) to the effects on the rate of diffusion. The present section refers to the effects when diffusion is very slow and the effects measured are those on the partition between the solution and the adsorbed forms. The temperature terms which are relevant are therefore those of Equations (2) and (8) of Table A8.1. Consider first Equation (8). This equation shows that increasing the temperature decreases the value of the binding constant. This means that, if the solution concentration is kept constant, high temperatures decrease adsorption. Alternatively, a higher concentration of the adsorbing material is required in solution in order to maintain the same amount of adsorption. However these effects may be modified by the effects of temperature on Equation (2). In this case the direction of the effect depends on both the sign of the value for potential and the sign of the charge on the adsorbing ion. For example, if an anion is adsorbing, and the potential is positive, an increase in temperature decreases the value of the term for potential. This therefore augments the effect due to Equation (8). Alternatively, if the potential is negative, an increase in the temperature increases the magnitude of the term and therefore opposes the effects due to Equation (8). Figure A8.6 shows that the model reproduces the direction of the effect for phosphate. Similar results have been obtained for molybdate and for zinc (Barrow 1986a,b).

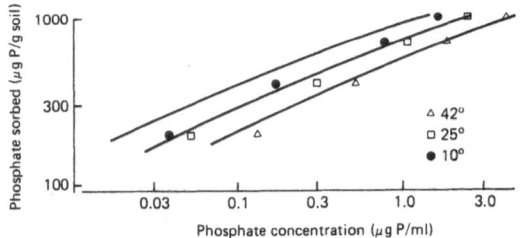

Fig. A8.6 Observed and modelled effect of changing the temperature at which the null-point phosphate concentration was measured after a long period of incubation. (From Barrow 1983a)

7) *Repeated additions behave as separate pulses*

It has been observed that repeated additions of a nutrient behave as separate pulses. This behaviour was summarised by Equation A7.7. For convenience, this equation is reproduced here:

$$ac^{b_1} = s\,t^{-b_2} + s_1\,(t-t_1)^{-b_2} + s_2\,(t-t_2)^{-b_2} \qquad\qquad A7.7$$

where an amount s is sorbed at $t = 0$ and an extra amount s_1 is added at t_1, and s_2 at t_2. The similarity between this equation and Equation A8.2 is apparent. That is, the observed behaviour of separate pulses is in accord with a diffusion mechanism acting as the rate-limiting step in the sorption process.

As noted in Chapter A7, Equation A7.7 is only an approximation to the behaviour. This equation must be modified because each subsequent addition is to a different surface — one for which the buffering capacity has been decreased because of the prior reaction. This effect is reproduced within the model by the parameter m_2. This parameter controls the change in surface electric potential as a result of diffusive penetration. The greater the change in potential, the greater the reduction in the subsequent buffering capacity. The effects of some differing values of this term are shown in Fig. A8.7. When the values for m_2 are large enough, there is a marked decrease in the subsequent buffering capacity. However the model does not reproduce the observed shape in the relation (Fig. A7.5).

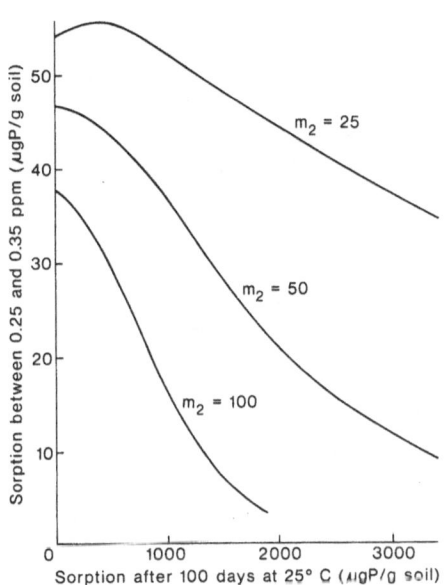

Fig. A8.7 Effects of varying the value of the parameter m_2 on the modelled buffering capacity of the soil when further additions of phosphate are made. The buffering capacity is indicated here by the difference in sorption between concentrations of 0.25 and 0.35 p.p.m. The modelled initial reaction was at 25°C for 100 days.

8) Desorption can be treated as repeated (negative) additions

Chapter A7 also shows that desorption can be treated as the opposite of a repeated addition and thus similarly treated as a separate pulse. This is also in accord with a solid state diffusion mechanism acting as the rate limiting step in the desorption process. Figure A8.8 shows that the model was able to reproduce the observed behavior of fluoride desorption. Note that, in this case, the parameter values that were used were those that had been found to describe the rate of fluoride sorption. That is, the model was used to "predict" fluoride desorption from the observed values for sorption. Note also that the model included the first stages of desorption when the reaction was dominated by the movement of adsorbed fluoride to the solution and the rate was not yet limited by the

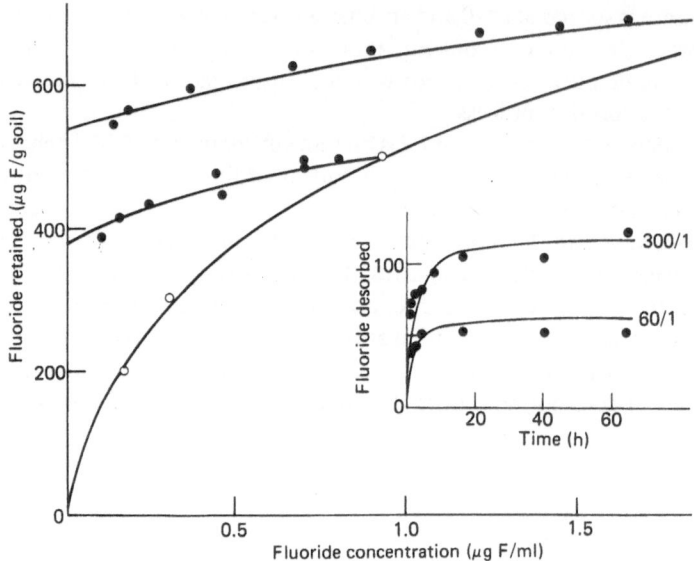

Fig. A8.8 Observed and modelled values for the desorption of fluoride that had been incubated with the soil for 4 days at 80°C. The inset shows modelled and observed values for two solution/soil ratios. (From Barrow 1986a)

diffusion step. The model provides a simple explanation of the observation that the longer the prior sorption phase, the slower the subseqent desorption phase. It is that more of the adsorbed material has penetrated the surface and that it has penetrated more deeply.

The model has also been shown to reproduce the main features of desorption for phosphate and for zinc (Barrow 1983b,1986b).

9) There are characteristic effects of pH for each nutrient

Chapter A5 showed that, for oxides, the four plane model provided a simple and consistent means of explaining the effects of pH on the adsorption of both anions and cations. It assumed that the effects were due to the interplay of two effects of pH. One of these was the effect on the electric potential of the surface, and the other was the effect on the species of ions present in the solution. This is broadly the explanation used when the model is applied to soil. At this level, the model may be used to explain the broad differences between the reactants. Thus sulfate adsorption is marked at low pH because the acid is already fully dissociated and the only effect observed is that due to the increasingly unfavourable surface potential as the pH is increased. In contrast, borate adsorption increases with increasing pH because the increasing proportions of the monovalent $H_4BO_4^-$ ion more than offset the decrease in electric potential of the surface.

At a more detailed level, the model may be tested against the observed effects of pH for a given reactant in order to find how closely it can reproduce them. As it is argued that we do not know from first principles the way the potential changes with pH, fitting the model to the data involves finding smooth curves which relate potential to pH and which then describe the observed effects on sorption. Figure A7.11 shows the outcome of this process for phosphate. The somewhat differing effects of pH on the individual soils were successfully described by assuming slightly different curves to relate potential to pH (Barrow 1984). In the presence of soil, the species of ions present may be affected by complexes formed with soil constituents. Thus for fluoride, sorption decreases at low pH because of the formation of complexes with aluminium (Chapter A7). This is consistent with the argument of the model that sorption depends, in part, on the fraction of the reactant present in solution as the adsorbing species. When fluoride sorption was expressed on the basis of equal concentrations of fluoride ion, sorption decreased with increasing pH (Fig A8.9). This is consistent with the general predictions of the model and,

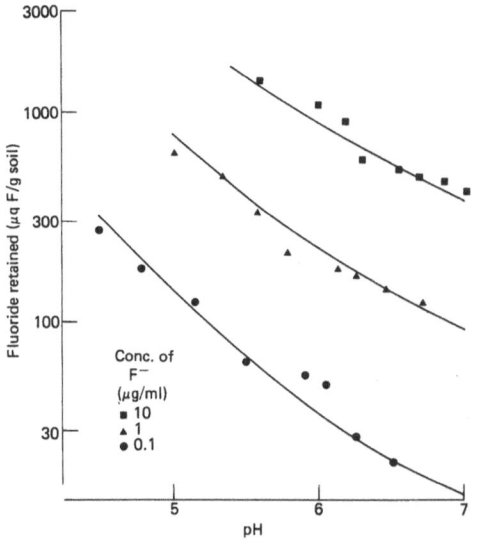

Fig. A.8.9   Observed and modelled effects of soil pH on the sorption of fluoride by soil at the indicated values for the F^- ion in solution. (From Barrow and Ellis 1986a)

furthermore, the effects were closely reproduced by the model.

For hydrolysable cations, the application of the model is a little more complex. If we wish to split the effects due to the changing proportions of the ions from those due to the changing electric potential, it is logical to plot sorption against the concentration of the presumed reacting ion. The remaining effect is then due to the effects of pH on the surface potential. However, for metals (for which it is assumed that the adsorbing ion is the $MeOH^+$ species) such plots often fall near a common line. This is illustrated for zinc in Figure A8.10. In this figure, the upper portion shows the effect of the indicated lime or acid treatments on the pH and on zinc sorption. The lower portion shows, that apart from the highest pH values, the points clustered about a common line when plotted against $ZnOH^+$ concentration. Similar results were reported for zinc in a range of soils by Barrow (1986c). It was also shown by Barrow (1985) that the data of Harter (1983) for the adsorption of lead, copper, zinc, and nickel (Fig A7.11) behaved similarly. Thus the model appears to require two mutually inconsistent hypotheses — that reaction of the cations is with variable charge materials but that the electric potential does not vary with pH. However it would also be argued that the metal ions would tend to react preferentially with the most-negative end of the spectrum of electric potentials. It has been suggested that such negative

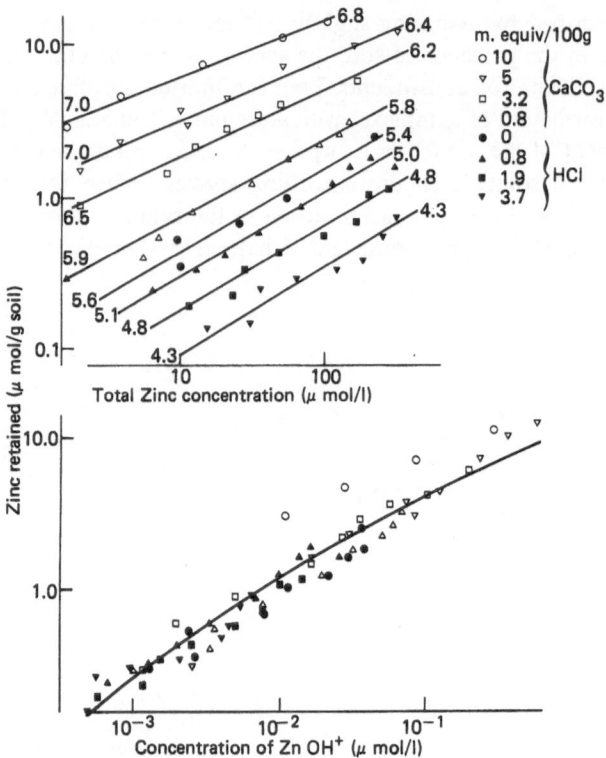

Fig. A8.10 Effect of plotting the data for the effect of pH on zinc sorption against the calculated concentration of the ZnOH$^+$ ion. (From Barrow 1986c)

potentials could arise from close association of the variable charge material with negative charge (Chapter A4). This could be as a result of silicate and phosphate present in the crystals and/or close association with negatively charged clays. It was shown in Chapter A4 that once the negative charge becomes large enough for the surface to be negative at moderate pH, the electric potential of the surface is indeed independent of pH. Thus this observation need not be inconsistent with the model. The critical question is whether the metals do indeed react with negative surfaces. If they do, then one would expect that increasing the concentration of salt in the medium would decrease sorption (Chapter A5). This is in fact the case (Fig. A8.11). This figure shows not only that increasing the concentration of salt decreased zinc sorption but also that the lines were close to parallel. This supports the argument that the electric potential of the sorbing surfaces did not change with pH. Thus the effects of salt concentration, and the interactions of this with pH are critical to understanding the system. These are considered further in the next section.

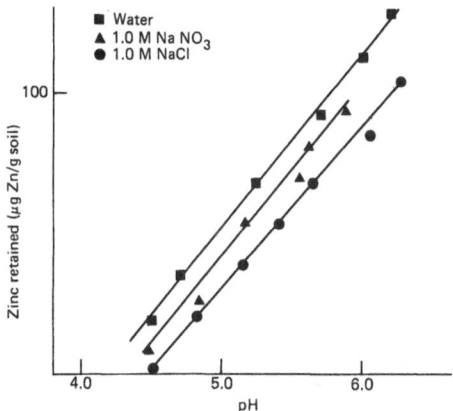

Fig. A8.11 Effect of pH and of salt concentration on the
sorption of zinc at a solution concentration of 1 p.p.m.
Zn. (From Barrow and Ellis 1986b)

10) *Interactions between pH and salt concentration*

It is an important observation that the effects of pH on phosphate sorption depend on the
concentration of sodium chloride in the background electrolyte (Fig. A7.13). The simplest
explanation (indeed perhaps the only satisfactory explanation) is that the interaction
between pH and electrolyte concentration occurs via the effects of electrolyte on the
electric potential of the surface (Chapter A5). This observation therefore supports the
general argument that reaction is with variable charge surfaces. The questions that then
arise are: what further conclusions can be drawn from these observations and are they
consistent with the model? As shown in Chapter A5, if an increase in the salt concentration
increases anion sorption, then the reacting surfaces are, on the average, negatively charged.
The converse applies if the increase in salt concentration decreases sorption. Thus, for the
soil illustrated in Fig. A7.13, the surfaces that have reacted with phosphate are, on the
average, negatively charged above about pH 5 and positively charged below about pH 5.
However, for the soil as a whole, the point of zero charge is at about pH 4.3 (Fig. A7.14).
Thus, at pH 4.3, phosphate has reacted with a subset of the total charged sites that is
positively charged. It is only when the pH is raised to about 5, that the subset of the sites
that have reacted with phosphate have an average charge of zero. It is a stringent test of the
model that it was able to reproduce the observations of Fig. A7.13 in a way that is
consistent with these deductions about the charge on the reacting surfaces. Figure A8.12
shows that at pH 4.3, phosphate was modelled as reacting with the positive tail of a
distribution centered on zero. At pH 5, the sites on which phosphate was sorbed were
modelled as having an average potential of zero, whereas for the total distribution, the
average potential was negative.

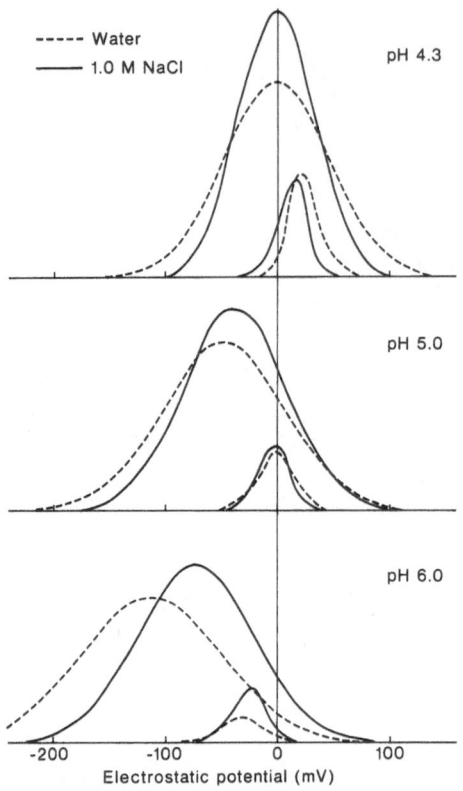

Fig. A8.12 Representation of part of the model used to describe the effects of pH and of salt concentration on phosphate sorption as shown in Fig. A7.13. The assumed distributions of potentials are shown for three pH values and for two salt concentrations. At each pH, the larger curves represent the distribution of total sites before the addition of phosphate. The areas under these curves are the same. The smaller curves represent the distribution of the phosphated sites at a solution concentration of 3.1 ppm of P. (From Barrow and Ellis 1986b)

The effects of the parameters of the model

The above section shows that the model is comprehensive. It is able to describe all of the observed characteristics of the reaction of anions and cations with soils. In most cases this description is quantitative. This section is therefore devoted to exploring the effects of the parameters of the model. It is thus another (albeit small) "knob—twisting" exercise.

The parameters of the model are indicated in the equations of Table A8.1. They are: the maximum adsorption; the binding constant K_i; the rate constants k_1 and k_2; the mean value of the inital potential ψ; the standard deviation of the initial potential σ; the coefficient to describe diffusion $\tilde{D}$; the feed-back coefficients m_1 and m_2; and the activation energy for diffusion E. The two further terms (A and B of Equations (7) and (8)) are effectively fixed once the values are allocated to $\tilde{D}$, E and K_i.

The term for maximum adsorption does not appear specifically in Table A8.1. It is however implied by the term θ— which is the fraction of the maximum adsorption for each component. The term for maximum adsorption can be considered as a scaling factor for the vertical axis of graphs for sorption. This is indicated in Fig A8.13 in which sorption is expressed as a proportion of the maximum adsorption. Changing the value of the term for maximum adsorption has the same effect as changing the scale for sorption. This is because maximum adsorption is treated in the model as a linear function of surface area.

96

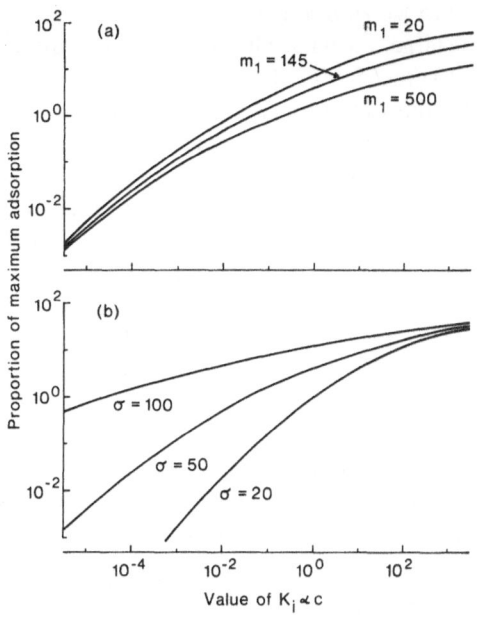

Fig. A8.13 Effect of varying the value of the parameters m_1 and σ on the modelled sorption. The other parameters were held constant at the values indicated in the program THETA.

The binding constant appears in the equations as a product with concentration. Increasing the binding constant therefore has the same effect as increasing the concentration. To emphasise this, the horizontal axis of Fig A8.13 is drawn as the product of concentration and binding constant. (The value for the fraction dissociated - α - is also included on this scale because varying it has the same effect as varying the other terms.)

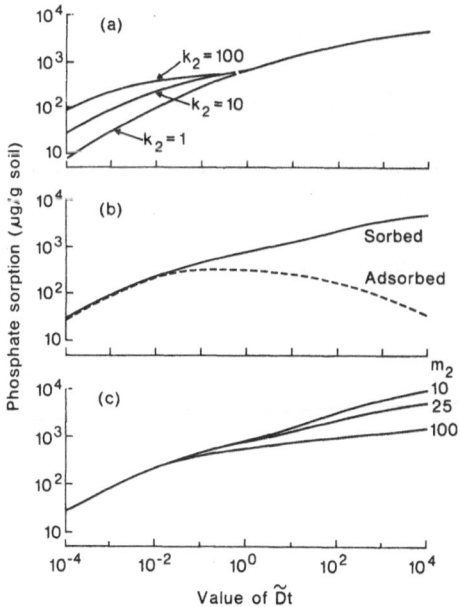

Fig. A8.14 Effect of varying the parameters k_2 and m_2 on the modelled effect of time on sorption. The other parameters were held at the values indicated in the program DIFPLUS. The centre part of the figure shows the modelled partition between the sorbed and the adsorbed forms when the parameters were held at the values indicated in the program.

The ratio of the rate terms k_1/k_2 is equal to K_i. Hence, if K_i is fixed, k_1 and k_2 vary together and only one need be specified. The larger the value allocated, the more quickly the initial adsorption reaction is completed (Fig A8.14a). This figure also shows that subsequent effects of time are then due to the diffusion process and are unaffected.

The term for electric potential occurs in the equations as a product with the term K_i. Changing the mean value of the initial potential therefore has similar effects to changing the value of K_i. The effects differ in that they operate through an exponential term and are therfore not linear but nevertheless also have the effect of changing the scale of the concentration axis.

The major effect of the term for the standard deviation of the distribution of potentials (σ) is on the shape of plots of sorption against concentration. Figure A8.13b shows that this term affects both the slope and the curvature of the logarithmic plots. When σ is large, the slope of the plots is low. Although the curvature is obvious over the nine log. cycles shown in this figure it would not be so apparent over the two to three log. cycles investigated in most experiments. When σ is smaller, the curvature is more marked and at low values for "concentration" the slope of the log. plots approaches unity. This means that at these concentrations a plot on a linear scale would be close to linear.

The coefficient to describe diffusion, $\tilde{D}$, occurs in the equation as a product with time. Over that part of the process for which diffusion is the rate-limiting step, increasing the value of $\tilde{D}$ therefore has the same effect as increasing the time. To indicate this, the horizontal scale of Fig.A8.13 has been drawn as the product of time and $\tilde{D}$ In this graph, changing the value of $\tilde{D}$ would have little effect on the first part of the process for which adsorption is rate limiting.

Most of the effect of the feed-back coefficient, m_1, is on the shape of plots of sorption against concentration. Figure A8.13a shows that much of the effect is at high values for concentration. Because large values for m_1 produce a bigger feed-back effect on the electric potential of the surface, they decrease the modelled value for sorption. The parameter m_1, together with the parameter σ, provide the main means of manipulating the shape of plots of sorption against concentration.

The other feed-back coefficient, m_2, reflects the effect of penetrated ions on the surface electric potential and thus on the surface concentration of truly adsorbed ions. Figure A8.14b shows that the modelled value for the adsorbed component reaches a maximum fairly early. It subsequently declines as the surface becomes more unfavourable. This declining concentration serves as a declining source for the diffusion process. Hence the larger the value for m_2, the slower the increase in total sorption (Fig. A8.14c). These effects of m_2 are only apparent on those parts of the process for which diffusion is the rate-limiting step. Thus the parameters m_2 and k_2 (and k_1) provide the main means of manipulating the shape of plots of sorption against time.

The activation energy, E, determines the extent to which an increase in the temperature increase the rate of diffusion. A common value for E is about 80kJ per mole. At this value, an increase in temperature of 10 degrees increases the rate about three-fold. Thus, for example, three days at $15°C$ would have about the same effect as one day at $25°C$. This approximation may be extended to a wider range of temperatures so that a change of temperature of say $30°C$ induces about a 27-fold change in rate — and is therefore very effective in "compressing" time. By substituting apropriate values for temperature and for

E in Equation(7) of Table A8.1, the relative rates for given changes of temperatures may be calculated more accurately than by this simple rule-of-thumb.

Interrelations between the parameters

The parameters that describe the behaviour of given sorbates are probably not independent. For example, ions that have a strong affinity for protons may also have a strong affinity for the electrophilic ions of the variable-charge oxides. This means that ions for which the degree of dissociation (α) is low tend to have a large value for K_i. Thus, for example, for sulfate, values of α are unity at soil pH and we may speculate that K_i may be about 10^2. On the other hand, for phosphate at pH 6, α is about 10^{-1} but K_i is about 10^6. The produce $K_i\alpha$ is thus about 10^3 times greater for phosphate. Hence phosphate is adsorbed more strongly — adsorption occurs at a lower concentration in solution and can occur despite an unfavourable change in the adsorbing surface and thus can occur at a higher pH. It also seems feasible that the diffusion term $\tilde{D}$ would be correlated with K_i since both reflect the ability of the ion to react with atoms of the substrate. Thus the continuing reaction is much more obvious for phosphate than for sulfate. It is hoped that further research will test the applicability of the ideas to other ions.

Summary

This Chapter shows that the ten characteristics of the reaction of soils with specifically sorbed materials can be reproduced by a model that is consistent with the reaction of models of the reaction with variable charge oxides.

References

Barrow, N.J. 1974. On the displacement of adsorbed anions from soil: 1. Dispacement of molybdate by phosphate and by hydroxide. Soil Science 116, 423-431.

Barrow, N.J. 1983a. A mechanistic model for describing the sorption and desorption of phosphate by soil. Journal of Soil Science 34,733-750.

Barrow, N.J. 1983b. On the reversibility of phosphate sorption by soils. Journal of Soil Science 34, 751-758.

Barrow, N.J. 1984. Modelling the effects of pH on phosphate sorption by soils. Journal of Soil Science 35, 283-297.

Barrow, N.J. 1985. Reaction of anions and cations with variable-charge soils. Advances in Agronomy 38, 183-230.

Barrow, N.J. 1986a. Testing a mechanistic model. I. The effects of time and temperature on the reaction of fluoride and molybdate with a soil. Journal of Soil Science, 37, 267-277.

Barrow, N.J. 1986b. Testing a mechanistic model. II. The effects of time and temperature on the reaction of zinc with a soil. Journal of Soil Science, 37, 287-295.

Barrow, N.J. 1986c. Testing a mechanistic model. IV.Describing the effects of pH on zinc retention by soils. Journal of Soil Science. 37, 295-302.

Barrow, N.J. and Ellis A.S. 1986a. Testing a mechanistic model. III. The effects of pH on fluoride retention by a soil. Journal of Soil Science. 37, 287-295.

Barrow, N.J. and Ellis A.S. 1986b. Testing a mechanistic model. V. The points of zero salt effect for phosphate retention and for acid/alkali titration of a soil. Journal of Soil Science 37, 303-310.

Halsey, G.D. 1952. The role of surface heterogeneity in adsorption. Advances in Catalysis. 4, 259-269.

Halsey, R.D. and Taylor H. S. 1947. The adsorption of hydrogen on tungsten powders. Journal of Chemical Physics. 15, 624-630.

Harter, R.D. 1983. Effect of soil pH on adsorption of lead, copper, zinc and nickel. Soil Science Society of America Journal. 47, 47-51.

Posner, A. M. and Barrow, N. J. 1982. Simplification of a model for ion adsorption on oxide surfaces. Journal of Soil Science. 33, 211-217.

Sposito, G. 1980. Derivation of the Freundlich equation for ion exchange reaction in soils. Soil Science Society of America Journal. 44, 652-654.

Van Riemsdijk, W.H., Boumans, L.J.M. and De Haan, F.A.M. 1984. Phosphate sorption by soils. I. A diffusion precipitation model for the reaction of phosphate with metal oxides in soil. Soil Science Society of American Journal. 48, 541-544.

Zeldowitsh, J. 1935. On the theory of the Freundlich adsorptiom isotherm. Acta Physico-chemica U.R.S.S. 1,961-974.

100

Chapter B1

Speciation in Solution

Calculation of the ion species present in solutions provides some simple examples of computer programs in which iteration methods sometimes have to be used to find the solution to equations. Consider, firstly, the simple case of the dissociation of a dibasic acid:

$$H_2A \rightleftharpoons + H^+ \tag{B1.1}$$

$$HA^- \rightleftharpoons A^{2-} + H^+ \tag{B2.2}$$

Then

$$K_{A1} = \frac{[HA^-][H]^+}{[H_2A]} \tag{B1.3}$$

and

$$K_{A2} = \frac{[A^{2-}][H^+]}{[HA^+]} \tag{B1.4}$$

where square brackets indicate concentrations. In this example, the H_2A might represent an acid such as H_2MoO_4 or it might represent a divalent metal ion surrounded by a sheath of water molecules. The divalent metal ion (eg Zn^{++}) can dissociate two protons from the water molecules to give, in effect, $ZnOH^+$ and $Zn(OH)_2$.

We can calculate the proportion present as one of these species (say A^{2-}) by expressing the other species in terms of A^{2-} using the equilibrium expressions above.

Thus:

$$\frac{[A^{2-}]}{[H_2A] + [HA^-] + [A^{2-}]} = \frac{K_{A1}\,K_{A2}}{[H^+]^2 + K_{A1}\,[H^+] + K_{A1}\,K_{A2}} \tag{B1.5}$$

The same process can be used for both of the other species present and can be extended to polybasic acids.

Provided one knows the values of K_{A1} and K_{A2}, equation B1.5 can be easily programmed to calculate the species present. However, suppose the dissociation we are considering is indeed that of molybdic acid. Molybdates are involved in several polymerisation reactions. There is some uncertainty about the products of these reactions but we will follow the scheme of Baes and Mesmer (1976). To illustrate the problem, we will consider only the first reaction they specify:

$$7\,MoO_4^{2-} + 8H^+ \rightleftharpoons Mo_7O_{24}^{6-} + H_2O \tag{B1.6}$$

$$K_{Mo3} = \frac{[Mo_7\,O_{24}^{6-}]}{[MoO_4^{2-}]^7\,[H^+]^8} \tag{B1.7}$$

The effect of this polymer is to add an extra term to the denominator of equation (B1.5) to give:

$$\frac{K_{Mo1} \; K_{Mo2}}{[H^+]^2 + K_{Mo1}[H^+] + K_{Mo1} \, K_{Mo2} + K_{Mo1} \, K_{Mo2} \, K_{Mo3} \, [MoO_4^{2-}] \, [H^+]^8} \qquad (B1.8)$$

The several other polymers specified by Baes and Mesmer (1976) each add an extra term to the denominator. The effect of these terms is to decrease the proportion of free MoO_4^{2-} ions as the concentration increases and as the pH falls. However the presence of these terms on the right hand side makes the equation rather awkward — we need to know the concentration of molybdate ions before we can calculate the proportion of molybdate ions. This is where iteration procedures are used. We simply make an initial arbitrary guess of the proportion of the molybdate present as MoO_4^{2-} (α). We then use that value to calculate a second value of α using equation (B1.8) (plus the appropriate extra terms). If the first estimate of α is too high, the calculated value will be too low. We therefore repeat the process using a guess that is between these two values — the geometric mean turns out to be a fairly efficient choice. This process is repeated until the guessed value of α and the value calculated from the guess, agree within acceptable limits. Once the proportion present as MoO_4^{2-} has been calculated, the proportions of the other species can be easily found — for example from equations B1.3, 1.4 and 1.7 "MOLLY" at the end of this chapter presents the program.

A similar problem occurs when association occurs between some of the species present to form ion pairs. Consider, for example, a solution containing both phosphate and zinc ions. Then by reactions analogous to those of equation (B1.1) (B1.2), the following species would be formed: H_3PO_4, $H_2PO_4^-$, HPO_4^{2-}, Zn^{2+}, $ZnOH^+$, and $Zn(OH)_2$. (We will confine the problem to pH values well below 12 and so ignore the trivalent phosphate ion.) In addition, the following reactions will occur:

$$Zn^{2+} + H_2PO_4^- \rightleftarrows Zn \, H_2PO_4^+ \qquad (B1.9)$$

and

$$Zn^{2+} + HPO_4^{2-} \rightleftarrows Zn_1HPO_4 \qquad (B1.10)$$

Hence

$$K_{ZnP2} = \frac{[ZnH_2PO_4^+]}{[Zn^{2+}] \, [H_2PO_4^-]} \qquad (B1.11)$$

$$K_{ZnP2} = \frac{[ZnHPO_4]}{[Zn^{2+}] \, [HPO_4^{2-}]} \qquad (B1.12)$$

The phosphate species present and the zinc species present therefore also include ZnH_2PO_4 and $ZnHPO_4$. If we want to calculate the proportion of phosphate ions present as, say, $H_2PO_4^-$, the ratio will be:

$$\frac{H_2PO_4^-}{H_3PO_4 + H_2PO_4^- + HPO_4^{2-} + ZnH_2PO_4^+ + ZnHPO_4}$$

Thus, in order to calculate the phosphate species, we need to know the zinc species. However the denominator for the zinc species will similarly contain the Zn-P complexes and, in order to calculate these, we need to know the phosphate species. The solution to this "chicken-and-egg" problem is again to make an initial guess. If we guess the proportions of $H_2PO_4^-$ and HPO_4^{2-}, we can use these guesses to calculate the zinc species present. From these provisional values of the zinc species, and thus of the Zn-P complexes, a further estimate of the phosphate species can be made. This process of alternatively calculating the phosphate species and then the zinc species can be continued until estimates of a species which is common to both classifications agree to within an acceptable limit. "ZNOHP" presents such a program.

References

Baes, C.F. and Mesmer, R.E. 1976. The hydrolysis of cations. John Wiley and Sons. New York.

```
1    Rem           MOLLY.BAS



10    Print"THIS PGME CALCULATES BY ITERATION THE PROPN OF MOLYBDATE PRESENT"
20    Print"AS MoO4= AFTER ALLOWANCE IS MADE FOR THE   POLYMERS Mo7O24(6-)"
22    Print" & Mo7O23(OH)(5-)"


25    Rem NOTE THAT THERE ARE PROBLEMS IN CALCULATING SOME OF THE TERMS DUE
26    Rem TO THE SIZE OF THE NUMBERS. TO OVERCOME THIS, SQUARE ROOTS HAVE
27    Rem BEEN USED AND A SEQUENCE OF MULTIPLICATIONS USED IN 330 & 340.



28    Rem SPECIFY CONSTANTS AS IN BAES & MESMER. ADJUSTMENTS FOR IONIC
29    Rem STRENGTH ARE GIVEN THERE - IN THIS CASE 0.03 WAS USED.
30    F1=10^-3.777
40    F2=10^-4.214
50    F3=10^28.87 :Rem NOTE THIS IS THE SQUARE ROOT OF THE CONSTANT
60    F4=10^31.07 :Rem ALSO SQUARE ROOT


69   On Error Goto 2000
70   Input"CONC OF MO IN MICROMOLE    ",Conc : Molarconc=Conc*1E-06
80   Print"CONC IN PPM IS    ",Conc*0.09596
90   Input"pH                  ",pH
100   Hconc=10^(-pH)


105    Rem   CHOOSE AN ARBITRARY INITIAL VALUE OF ALPHA
110    Alpha=0.5


120    Gosub 250
130    Alpha=Alpha2
140    Gosub 250


145    Rem IF THE TWO ESTIMATES OF ALPHA ARE CLOSE ENOUGH, QUIT
150    Ratio=Diff/Alpha
160    If Abs(Ratio)<1E-03 Then 190


165    Rem TAKE GEOMETRIC MEAN AS NEW ESTIMATE
170    Alpha=Exp((Log(Alpha)+Log(Alpha2))/2)
180    Goto 140
```

```
185      Rem PRINT OUT RESULTS
190      Print : Print"VALUE OF ALPHA                    ",Alpha
200      Print"CONC OF MO4= IN MICROMOLES     ",Alpha*Conc
205      Print : Print"PROPORTIONS ARE: "
210      Print " UN-CHGD       MONO-V        DI-V        POLLYM1        POLLYM2"
220      Print  Using"##.#####     ";Term1/Sum,Term2/Sum,Term3/Sum,Term4/Sum,Term5/Sum
230      Print : Print
240      Input"CONTINUE (Y/N)?   ", Answ$ : If Answ$="Y" Or Answ$="y" Goto 70
245      End


250      Rem CALCULATE ALPHA
260      Term1=Hconc^2.0
270      Term2=F1*Hconc
280      Term3=F1*F2
290      Prod=Alpha*Molarconc^3.0 :Rem USE THE SQUARE ROOT OF THE REQUIRED TERM

300      Rem IF pH IS ABOVE 7 THE POLYMER TERMS WILL HAVE DISAPPEARED
310      If Hconc<1E-07 Then Term4=0 : Term5=0 : Goto 350

320      Rem BREAK UP THE CALCULATION OF THE POLYMER TERMS TO AVOID OVERFLOWS
330      Term4=F1*F2*F3*Prod*F3*Prod*Hconc^4.0*Hconc^4.0
340      Term5=F1*F2*F4*Prod*F3*Prod*Hconc^4.0*Hconc^5.0

350      Sum=Term1+Term2+Term3+Term4+Term5
370      Alpha2=Term3/Sum
380      Diff=Alpha-Alpha2
390      Return


2000     Print "ERROR NO ",Err,"  AT LINE NO   ",Erl : Resume Next
```

```
1    Rem                    ZNOHP.BAS


10   Print"PROGRAM CALCULATES IONS PRESENT IN MIXTURES OF Zn & P SOLUTIONS"
15   Print"IT IS ASSUMED THAT THE BACKGROUND ELCTROLYTE IS   0.01M CALCIUM "
20   Print"NITRATE. FOR OTHER ELECTROLYTES THE CONSTANTS SHOULD BE CHANGED."
25   Print"CONSTANTS FOR ZINC FROM BAES & MESMER, FOR Zn-P COMPLEXES FROM"
26   Print"NRIAGU(1973) GEOCHIMICA ET COSMOCHIMICA ACTA 37, 2357-2361."


29   Rem SPECIFY CONSTANTS
30   F1=7.745E-10
40   F2=1.148E-08
50   F3=39.8
60   F4=1995.2
70   Fp1=0.011
80   Fp2=7.5E-08
90   Print"DISS CONSTANTS FOR Zn++ & ZnOH+",F1,F2
100  Print"ASSOCIATION CONST FOR ZnH2PO4+ ",F3
110  Print"ASSOCIATION CONST FOR ZnHPO4    ",F4
120  Print"DISSOCIATION CONSTANTS FOR P    ",Fp1,Fp2


130  Print : Print : Print
140  Print"NOTE INPUT CONC IS EXPECTED AS PPM. OUTPUT IS MICROMOLES."
150  Print "CHANGE IF INCONVENIENT."
180  Print : Print
190  Input"pH   ",pH
210  Input"TOTAL ZN CONC (ppm)        ",Znconcppm
220  Znconc=Znconcppm/65.3e3 :    Rem CHANGE TO MOLAR
230  Print"ZN MICROMOLES      ",Znconc*1e6
240  Input"P CONC.(ppm)             ",Pconcppm
250  Pconc=Pconcppm/31e3
260  Print"P MICROMOLES      ",Pconc*1e6


270  Rem  FIRST EST. OF P IONS - IGNORING ZN.
280  Hconc=10^(-pH)
290  Pdenom=Fp1*Fp2+Fp1*Hconc+Hconc^2.0+Fp1*Hconc*F3*Znconc
300  AlphamonoP=Hconc*Fp1/Pdenom
310  AlphadiP=Fp1*Fp2/Pdenom
315  Diff=1

317    Rem START CYCLING
320    While Abs(Diff)>0.001
```

106

```
325    Znpterm1=Hconc^2.0*F3*AlphamonoP*Pconc
330    Znpterm2=Hconc^2.0*F4*AlphadiP*Pconc
335    Zndenom=Hconc^2.0 + Hconc*F1 + F1*F2 + Znpterm1 + Znpterm2
340    ConcdiZn=Znconc*Hconc^2.0/Zndenom
350    AlphaZnP=Znpterm1/Zndenom

355    Pznterm1=Fp1*Hconc*F3*ConcdiZn
360    Pznterm2=Fp1*Fp2*F4*ConcdiZn
365    Pdenom=Fp1*Fp2 + Fp1*Hconc+Hconc^2.0 + Pznterm1 + Pznterm2
370    AlphamonoP=Fp1*Hconc/Pdenom
380    AlphadiP=Fp1*Fp2/Pdenom
390    AlphaZnP2=Pznterm1/Pdenom


400    ConcZnP=AlphaZnP*Znconc
410    ConcZnP2=AlphaZnP2*Pconc
430    Diff=(ConcZnP-ConcZnP2)/ConcZnP

440    Wend

460    ZnOHconc=Znconc*Hconc*F1/Zndenom


470    Rem   PRINT OUT ANSWERS
480    Print Using"CONC OF ZNOH+    ####.##### MICROMOLES";ZnOHconc*1E6
500    Print Using"CONC OF ZnH2PO4+ ####.##### MICROMOLES";AlphaZnP*Znconc*1E6
510    ZndiP=Znconc*Hconc^2.0*F4*AlphadiP*Pconc/Zndenom
520    Print Using"CONC OF ZnHPO4   ####.##### MICROMOLES";ZndiP
530    Print
540    Print"IN PPM THE ZN IS (AS ZN++, ZNOH+, Zn1P, Zn2P) "
550    Ratio=65.3e3
560    Print ConcdiZn*Ratio,ZnOHconc*Ratio,AlphaZnP*Znconc*Ratio,ZndiP*Ratio
561    Print
562    Print"IN FRACTIONS THE ZN IS....."
565    Print ConcdiZn/Znconc,ZnOHconc/Znconc,AlphaZnP,ZndiP/Znconc
570    Ratio2=31.0e3
580    Print
590    Print"IN PPM THE P IS (AS H2PO4-, HPO4=, ZN1P,ZN2P)"
600    Print AlphamonoP*Pconc*Ratio2, AlphadiP*Pconc*Ratio2,
605    Print AlphaZnP*Znconc*Ratio2,ZndiP*Ratio2
610    Print : Print

620    Input "SHALL WE CONTINUE (Y/N) ? ",Answ$
630    If Answ$="Y" Or Answ$="y" Goto 190
640    End
```

Chapter B2

Solving simultaneous equations — the four-layer model

The previous chapter presented some simple iterative programs. These involve guessing a value of a property — e.g. α — and then using that value to calculate a second value. Thus the guess, and the calculation based in it, are two estimates of the value of the same property. In such cases it is fairly easy to refine the guess by simply comparing the two values — in MOLLY by taking the geometric mean. However there are some cases in which a different approach is needed.

Consider the four-layer model described in Chapters A3 and A5. The model postulates that the ions can be pictured as being arranged in four layers of a given mean electric potential. A set of equations describes the charge and the electric potential in the layers. Solving the equations involves allocating values to eight variables — four values of potential and four values of charge. As there are eight equations (Table B2.1), the

Table B2.1

The sequence of equations used for the four-plane model

Step	Equations	
1		Estimate ψ_s*
2	1	$\sigma_S = \dfrac{Ns\,[K_H\,a_H\,\exp(-F\,\psi/RT) - K_{OH}\,\exp(F\,\psi_a/RT)]}{1 + K_H\,a_H\,\exp(-F\,\psi/RT) + K_{OH}\,a_{OH}\,\exp(F\psi_a/RT)}$
3	2	$\psi_a = \psi_s - \sigma_s/G_{sa}$
4	3	$\sigma_a = \dfrac{N_T\,z_i\,K_i\,a_i\,\exp(-z\,F\,\psi_a/RT)}{1 + K_i\,a_i\,\exp(-z\,F\psi_a/RT)}$
5	4	$\psi_\beta = \psi_a - (\sigma_s + \sigma_a)/G_{a\beta}$
6	5	$\sigma_\beta = \dfrac{Ns\,[K_{cat}\,[Cat]\,\exp(-z_{cat}\,F\,\psi_\beta/RT) - K_{an}\,[An]\,\exp(z_{an}\,F\,\psi/RT]}{1 + K_{cat}\,[Cat]\,\exp(-z_{cat}\,F\,\psi_\beta/RT) + K_{an}\,[An]\,\exp(z_{an}\,F\,\psi/RT)}$
7	6	$\psi_d + \psi_\beta - (\sigma_s + \sigma_a + \sigma_\beta)/G_{\beta d}$
8	7	$\sigma_d = -1.22 \times 10^{-10}\,C^{\,0.5}\,\sinh(0.0195\,z_{cat}\,\psi_d)$
9	8	$\sigma_s + \sigma_a + \sigma_\beta + \sigma_d = 0$

* The symbols: ψ indicates electrical potential, σ charge and G capacitance, the subscripts s, a, β and d indicate the successive mean planes of adsorption for ions near the surface, N_T and N_S are maxima for ion adsorption, K indicates binding constants as indicated by the subscripts and F,R and T have their usual meaning.

equations must be soluble. Again the method adopted was to make an initial guess of the value of one of the variables. The most convenient variable for this is the electric potential of the innermost layer (ψ_S). The Nernst equation is used for a first estimate. (Especially where adsorption occurs, this estimate can be appreciably different from the final value.) Given a value for ψ_S, the first seven equations in Table B2.1 may be solved successively. At this stage, values will have been allocated to all of the variables. If these values were the correct ones, the eighth equation would be true and all the charges would add to zero — that is, the residue will be zero. This is most unlikely and so we have to try other values of ψ_S to find the correct one. Two approaches are effective. One is a stepwise procedure. We simply take a step — that is, we change the value of ψ_S. If the residue decreases we assume we are going in the correct direction and so take another step. Eventually a point will be reached at which the sign of the residue changes and we will know we have gone too far. We therefore reverse the direction of the steps but make them smaller. This process is continued until sufficient precision has been reached. The stepwise procedure is fairly robust and, provided there are no reverses in the direction of the relation between the residue and the parameter being altered, it will eventually solve the problem. However it can be fairly slow. If the relation between the residue and the parameter is not too curved, an extrapolation procedure is more efficient. For the first two cycles, this procedure is the same as before — that is, a first estimate is made and then a second estimate, a step removed from the first estimate. However the third step is made by constructing a chord to the curve (Fig. B2.1) and extrapolating this to zero. If the relation between the residue and the parameter were linear, only one extrapolation would be needed. More usually it is curved and so the extrapolation will miss the zero point (Fig. B2.1). However a new extrapolation can now be made based on the most recent estimate, and on the previous one. This will be closer to the zero point. Repetition for two or three more cycles will usually approach the correct value closely.

For the four-layer model, the extrapolation method only works well if the outer capacitance is large. The more robust approach is therefore to use the step-wise procedure to approach the end point and then to change to the extrapolation procedure in order to find the end-point quickly.

A program describing four-layer model follows. It is constructed so that a diagrammatic representation of the model appears on the screen — thus helping the operator to identify the parameters. The model requires the equilibrium concentration of adsorbate. Some generality is provided in the calculation of the ion species present but, for example, there is no provision for ions such as molybdate which may polymerise. To use the model for molybdate, it would be necessary to incorporate MOLLY.

For this model — and for others which follow — some suggested values of the parameters are indicated. It is important to stress that these are not "correct" values. They are merely rounded values which have been found to be useful in some typical applications. Thus, in the present case the suggested values may model fairly closely the adsorption of phosphate on goethite. They are provided in order to provide a simple starting point. For a particular application, however, it will probably be necessary to modify the parameters. The effects of these modifications are given in Chapter A5.

When the outer capacitance is small, the model is very sensitive — very small changes in the value of the potential of the inner layer lead to appreciable differences in the charges. It may therefore be desirable to work with double precision if this is available on your computer. Alternately, it may be necessary to decrease the value of the "Difference" (statement 1005 in the accompanying program) to permit the cycle to finish.

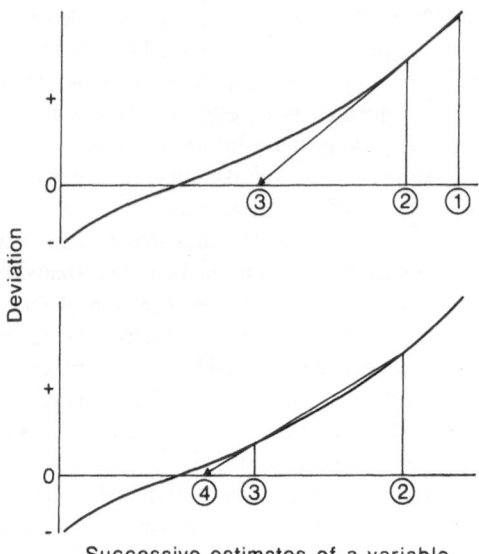

Successive estimates of a variable

Fig. B2.1 Diagramatic representation of the successive extrapolation method for finding the simultaneous solution to a set of equations. In the present case, the variable estimated is the electrical potential in the innermost plane and the deviation measured is that for equation 8 of Table B2.1

```
1     Rem                    BOWDEN.BAS


10    Print "THIS PROGRAM IS THE EXTENDED BOWDEN MODEL FOR ADSORPTION."
15    Lprint "THIS PROGRAM IS THE EXTENDED BOWDEN MODEL FOR ADSORPTION."
16    Lprint:Lprint
20    Print "THERE IS PROVISION FOR NON-SYMETRICAL ELECTROLYTES & IT CAN BE"
30    Print "USED  EITHER IN THE PRESENCE OR THE  ABSENCE  OF SPECIFIC"
35    Print "ADSORPTION."
40    Print "THERMODYNAMIC DISSOCIATION CONSTANTS FOR  P  ARE SUPPLIED."
42    Print "TO WORK WITH OTHER NUTRIENTS THE CONSTANTS MUST BE SUPPLIED."
45    Print "SOME SUGGESTED (ROUNDED) VALUES OF THE PARAMETERS ARE GIVEN"
47    Print "-IN BRACKETS"


50    Rem FOR SIMPLER MODELS SET Kcat=0, Kan=0 & / OR SET APPROPRIATE VALUES
55    Rem OF CAPACITANCE (Gsa,Gac,Gcd) LARGE - E.G. 1e10
60    Rem SYMBOLS FOR CHARGE, POTENTIAL, & CAPACITANCE ARE DEFINED IN THE
65    Rem INITIAL OUTPUT.
70    Rem Kh IS K(H), Koh IS K(OH), Kcat IS K(Cation), Kan IS K(Anion),
90    Rem Kdiss1,Kdiss2,& Kdiss3 ARE DISSOCIATION CONST OF PHOSPHORIC ACID
100   Rem OR OF OTHER ACID AS ENTERED
110   Rem Kaff IS AN ARRAY OF AFFINITY COEFF FOR MONO, DI, & TRIVALENT IONS,
115   Rem M IS AN ARRAY OF CONCS OF THESE IONS.
120   Rem Ads & Chg ARE ARRAYS OF TERMS IN THE BOWDEN ADSORPTION CHARGE EQN.
130   Rem Adsion IS AN ARRAY OF ADSORPTION OF INDIVIDUAL ADSORBED SPECIES


140   Dim Act(3),Concion(3),Kaff(3),Chg(3),Ads(3),Adsion(3),Kdiss(3)
150   On Error Goto 2000


160   Rem THE NEXT STATEMENTS CONTROL INPUTS OF PARAMETERS


170   Print" MODEL IS REPRESENTED:"
180   Print"  SURFACE       ADSORBED      COUNTER        DIFFUSE"
190   Print"   PSI(S)        PSI(I)        PSI(C)         PSI(D)"
200   Print"   Psi1          Psi2          Psi3           Psi4"
210   For I=1 To 5
214   Print"     |             |             |             |"
216   Next I
220   Print"       | CAPTNCE Gsa | CAPTNCE Gac|  CAPTNCE Gcd |"
230   Print"  SIGMA(S)      SIGMA(I)      GAMMA(NA)      SIGMA(D)"
240   Print"                              GAMMA(CL)"
```

```
250    Print" Sigma1           Sigma2      Gamma3 & 4      Sigma4



260    Print : Print : Print : Print"VALUES OF PARAMETERS"
265    Index=0
270    Print"MAX ADSPTN IN S PLANE IN MICROMOL/SQ.M ( 10 )",Tab(55);
272    Input Maxadss
273    Lprint"MAX ADSPTN IN S PLANE IN MICROMOL/SQ.M ( 10 ),Tab(55), Maxadss
275    Print "CAPACITANCES TO BE ENTERED AS FARADS/SQ.M."
280    Print"CAPACITANCE Gsa( 3 )",Tab(55); : Input Gsa
285    Lprint"CAPACITANCE Gsa ",Tab(55), Gsa
290    Print"CAPACITANCE Gac( 5 )",Tab(55); : Input Gac
295    Lprint"CAPACITANCE Gac",Tab(55),Gac
300    Print"CAPACITANCE Gcd(V. LARGE eg 1e12 or 0.2 )",Tab(55); : Input Gcd
305    Lprint"CAPACITANCE Gcd",Tab(55),Gcd



310    Print"CHECK - CAPNCE FROM SURFACE TO COUNTER IONS IS";1/(1/Gsa+1/Gac)
312    Print:Print
315    Print"IF YOU ARE WORKING AT CONST IONIC STRENGTH YOU MAY WANT TO USE"
316    Print " CONCENTRATION NOT ACTIVITY.  IF SO,YOU WILL HAVE TO ENTER"
317    Print " DISSOCIATION CONSTS FOR THAT IONIC STRENGTH"



320    Input "CONCENTRATION OR ACTIVITY (C/A)?",Act$
325    If Act$="A" Or Act$="a" Then Lprint "                ACTIVITY      "
330    Print"K(H) ( 1e7 )",Tab(55);  Input Kh
331    Print"K(OH) ( 1e4 )",Tab(55); : Input Koh
333    Lprint"K(H) ",Tab(55),Kh : Lprint"K(OH)",Tab(55),Koh
335    Print"K(CAT) ( 1 )",Tab(55); : Input Kcat
336    Print"K(AN) ( 1 )",Tab(55); : Input Kan
337    Lprint"K(CAT)",Tab(55),Kcat : Lprint"K(AN)",Tab(55),Kan
338    Lprint:Lprint



340    Input "ARE SPECIFICALLY ADSORBING IONS PRESENT (Y/N)?",Adsrp$
342    If Adsrp$="N" Or Adsrp$="n" Goto 410
345    If Index>0 goto 440
350    Print"MAX. ADS IN PLANE A IN MICROMOLES PER SQ M.(FOR P 2.5 )",Tab(55);
352    Input Maxada
355    Lprint"MAX. ADS IN PLANE A IN MICROMOLES PER SQ M. ",Tab(55),Maxada
360    Kdiss(1)=7.58E-03 : Kdiss(2)=6.166E-08 : Kdiss(3)=2.137E-13
365    If Act$="C" Or Act$="c" Then Goto 385
```

```
370   Input "IS ADSORBATE PHOSPHATE (Y/N)? ",Phos$
372   If Phos$="Y" Or Phos$="y" Then 397
385   Print"DISS CONSTS FOR ADSORBING SPECIES. ENTER ZEROES FOR 2ND &/OR "
387   Print"3RD CONST IF THEY DO NOT EXIST. NOTE  CONSTANTS - NOT LOG CONST"
390   Lprint"DISS CONSTS FOR ADSORBING SPECIES. "
391   Print"FIRST",Tab(55); : Input Kdiss(1)
392   Lprint"FIRST",Tab(55),Kdiss(1)
393   Print"SECOND",Tab(55); : Input Kdiss(2)
394   Lprint"SECOND",Tab(55),Kdiss(2)
395   Print"THIRD",Tab(55); : Input Kdiss(3)
396   Lprint"THIRD",Tab(55), Kdiss(3)

397   Print"BINDING CONSTANTS FOR ADSORBING SPECIES. USE MICROMOLAR SCALE"
398   Print"ENTER ZEROES FOR NON ADSORBED SPECIES."
399   Lprint:Lprint"BINDING CONSTANTS FOR ADSORBING SPECIES.(L/MICROMOLE)"
400   Print"SUGGESTED VALUES ARE FOR P"
401   Print"MONOVALENT (0 )";Tab(55); : Input Kaff(1)
402   Lprint"MONOVALENT ";Tab(55), Kaff(1)
405   Print"DIVALENT ( 10) )";Tab(55); : Input Kaff(2)
406   Lprint"DIVALENT";Tab(55),Kaff(2)
407   Print"TRIVALENT ( 0 )";Tab(55); : Input Kaff(3)
408   Lprint"TRIVALENT";Tab(55), Kaff(3)
409   If Index>0 Goto 440

410   Print"VALENCY OF ELECTROLYTE CATION",Tab(55); : Input Vcat
411   Lprint:Lprint"VALENCY OF ELECTROLYTE CATION",Tab(55),Vcat
420   Print"VALENCY OF ELECTROLYTE ANION",Tab(55); : Input Van
421   Lprint"VALENCY OF ELECTROLYTE ANION",Tab(55),Van
430   Print : Print"  VALUES OF VARIABLES"
425   Lprint:Lprint
440   Print"pH",Tab(55); : Input Ph
442   Lprint"pH",Tab(55),Ph
445   Print"ELECTROLYTE CONC.(MOLAR SCALE)",Tab(55); : Input Concelec
446   Lprint"ELECTROLYTE CONC.(MOLAR SCALE)",Tab(55), Concelec

450   Rem SIMPLE CALCULATION OF ACTIVITY COEFF FOLLOWS. ALTER IF YOU WANT
451   Rem  MORE COMPLEX VERSION. THIS SIMPLE VERSION WILL NOT GIVE GOOD
452   Rem VALUES FOR HIGH ELECTROLYTE CONC.

460   Mu=0.5*(Van*Concelec*Vcat^2+Vcat*Concelec*Van^2)
```

```
470    Actcoef=10^(-0.5*Sqr(Mu)/(1.0+Sqr(Mu)))
485    If Act$="C" Or Act$="c" Then Actcoef=1
490    Rem IF USING THERMODYNAMIC DISS CONSTANTS, ADJUST FOR IONIC STRENGTH.
500    Kdiss1=Kdiss(1)/Actcoef^2 : Kdiss2=Kdiss(2)/Actcoef^4
505    Kdiss3=Kdiss(3)/Actcoef^6
510    If Adsrp$="Y" Or Adsrp$="y" Then 511 Else 515
511    Print"CONC. OF ADSORBATE IN MICROMOLES/L";Tab(55); : Input Concads
512    Lprint"CONC. OF ADSORBATE IN MICROMOLES/L";Tab(55),Concads
515    Print : Print


520    Rem THE NEXT STATEMENTS MAKE A FIRST ESTIMATE OF PSI(S).
530    Conch=10^(-Ph) : Concoh=1E-14/Conch
550    Term=Sqr(Koh*1E-14/Kh) : Zpc=-Log(Term)/Log(10)
560    Psi1=58*(Zpc-Ph)
570    Print Using"NERNST EST  ####.#";Psi1


580    Rem THE NEXT STATEMENTS ARE THE SET OF CHARGING EQUATIONS.
590    Rem THEY END AT 980 WITH A SECOND ESTIMATE OF SIGMA)(S) - Estsigma1

600    Print
610    Delta1=10:Ii=0
620    I=1
630    Hterm=Kh*Conch*Exp(-0.039*Psi1)
640    Ohterm=Koh*Concoh*Exp(0.039*Psi1)
641    Gammah=Maxadss*Hterm/(1+Hterm+Ohterm)
642    Gammaoh=Maxadss*Ohterm/(1+Hterm+Ohterm)
650    Sigma1=Gammah-Gammaoh
660    Psi2=Psi1-Sigma1*96.487/Gsa
661    Rem USE FARADAY CONSTANT TO GIVE POTNL IN MILLIVOLTS
670    If Concads=0 Then Sigma2=0 : Goto 770
680    T1=Conch^3+Conch^2*Kdiss1+Conch*Kdiss1*Kdiss2+Kdiss1*Kdiss2*Kdiss3
690    Concion(1)=Concads*Kdiss1*Conch^2/T1
700    Act(1)=Concion(1)*Actcoef
710    Concion(2)=Concads*Conch*Kdiss1*Kdiss2/T1
720    Act(2)=Concion(2)*Actcoef^4
730    Concion(3)=Concads*Kdiss1*Kdiss2*Kdiss3/T1
740    Act(3)=Concion(3)*Actcoef^9


745    Rem  THE NEXT SECTION CALCULATES THE CHARGE DUE TO ASORPTION
750    Sumchg=0 : Sumads=0
752      For K=1 To 3
755        Ads(K)=Kaff(K)*Act(K)*Exp(K*0.039*Psi2)
```

```
757    Chg(K)=-K*Ads(K)
759    Sumchg=Sumchg+Chg(K)
760    Sumads=Sumads+Ads(K)
765    Next K
767    Sigma2=Maxada*Sumchg/(1+Sumads)
770    Psi3=Psi2-(Sigma1+Sigma2)*96.487/Gac


800    Rem CALC MOLARITY OF INDIVIDUAL IONS FOR NON SYMMETRICAL ELECTROLYTE.
810    Conccat=Concelec : If Van>Vcat Then Conccat=Concelec*Van
820    Conccat=Conccat*Actcoef^(Vcat^2)
840    Concan=Concelec : If Vcat>Van Then Concan=Concelec*Vcat
850    Concan=Concan*Actcoef^(Van^2)
860    Anionterm=Kan*Concan*Exp(0.039*Van*Psi3)
865    Cationterm=Kcat*Conccat*Exp(-0.039*Vcat*Psi3)
880    Gamma4=Maxadss*Van*Anionterm/(1+Anionterm+Cationterm)
890    Gamma3=Maxadss*Vcat*Cationterm/(1+Anionterm+Cationterm)
900    Psi4=Psi3-(Sigma1+Sigma2+Gamma3-Gamma4)*96.487/Gcd
930    If Vcat=Van Then 931 Else 940
931    T5=Vcat*0.0195*Psi4
932    If Abs(T5)>87 Then Sigma4=-Sgn(Psi4)*1e35: Goto 980
934    T6=0.5*(Exp(T5)-Exp(-T5))
936    Sigma4=-1.22*Sqr(Concelec)*T6
938    Goto 980
940    X1=0.039*Psi4*Vcat : X2=0.039*Psi4*Van
945    If Abs(X1)>87 Then Sigma4=-Sgn(Psi4)*1e35:Goto 980
950    W3=(Exp(-X1)-1)/Vcat : W4=(Exp(X2)-1)/Van
960    W5=Sqr(W3+W4) : If Psi4>0 Then W5=-W5
970    Sigma4=.61*Sqr(Concelec)*W5
980    Estsigma1=-Sigma2-Gamma3+Gamma4-Sigma4

985    Rem   END OF CALCULATION OF CHARGING EQUATIONS


990    Rem   NOW COMPARE THE 2 ESTIMATES OF SIGMA(S) & ADJUST PSI(S)
1000   Difference=Sigma1-Estsigma1
1005   If Abs(Difference)<1E-04   Goto 1140
1006   Delta2=Sgn(Difference)*Delta1


1007   Rem IF DIFFERENCE LARGE USE INCREMENTAL SEARCH
1008   If Abs(Difference)>100 Goto 1110


1009   Rem WHEN DIFFERENCE SMALL ENOUGH USE LINEAR EXTRAPOLATION
1010   If I=1 Then Prepsi=Psi1 : Goto 1090
1020   Slope=(Difference-prediff)/(Prepsi-Psi1)
```

```
1030    Intercept=Difference+Slope*Psil
1035    Prepsi=Psil
1037    If Slope=0 then Psil=Prepsi*0.999: Goto 1090
1040    Psil=Intercept/Slope
1050    If Psil=Prepsi Then Psil=Psil*0.999
1090    Prediff=Difference
1091    If I=11 Then  1092 Else 1094
1092    Print "SLOW CONVERGENCE. PRINT OUT SUCCESSIVE ESTS OF  POTNL & OF"
1093    Print" DIFFERENCE"
1094    If I>10 Then Print I,Prepsi,Difference
1095    If I>20 And Abs(Difference)<1e-3 Goto 1140
1096    If I=30 Then 1097 Else 1100
1097    Print "EQUATIONS WON'T CONVERGE. PERHAPS Gcd IS TOO SMALL GIVING
1098    Print" HIGH SENSITIVITY TO CHANGES IN PSI1":GOTO 1270
1100    I=I+1
1105    If I=2 Then Psil=Psil+Delta2
1106    Goto 630
1108    Rem THIS IS THE END OF THE EXTRAPOLATION.


1109    Rem THIS IS THE BEGINNING OF THE ALTERNATE INCRENMENTAL SEARCH
1110    I=2:Ii=Ii+1
1115    If Ii=1 Then 1128
1125    Prod=Sgn(Difference)*Sgn(Prediff)
1126    If Prod<0 Then Delta1=0.49*Delta1
1128    Prepsi=Psil:Psil=Psil+Delta2
1129    Prediff=Difference
1130    Goto 630
1132    Rem END OF INCREMENTAL SEARCH


1140    Rem THE NEXT STATEMENTS CONTROL OUTPUT.
1146    Lprint:Lprint
1150    Print Tab(30);:Print" SURFACE        ADSORBED        COUNTER        DIFFUSE"
1160    Print"POTENTIAL (M.VOLTS)        ";
1170    Print Using"      ####.###";Psil,Psi2,Psi3,Psi4
1171    Print"CHARGE (MICRO M/SQ.M.)    ";
1172    Print Using"        ##.###";Sigma1,Sigma2,Gamma3-Gamma4,Sigma4
1175    Print Tab(23):Print Using"GAMMA H    ##.###";Gammah;
1176    Print Using"        GAMMA CAT. ##.###";Gamma3
1177    Print Tab(23):Print Using"GAMMA OH  ##.###";Gammaoh;
1178    Print Using"        GAMMA AN.  ##.###";Gamma4
1180    LprintTab(30);:Lprint" SURFACE        ADSORBED        COUNTER        DIFFUSE"
1181    Lprint"POTENTIAL (M.VOLTS)        ";
1182    Lprint Using"      ####.###";Psil,Psi2,Psi3,Psi4
1183    Lprint"CHARGE (MICRO M/SQ.M.)    ";
```

```
1184    Lprint Using"         ##.###";Sigma1,Sigma2,Gamma3-Gamma4,Sigma4
1185    Lprint Tab(23):Lprint Using"GAMMA H   ##.###";Gammah;
1186    Lprint Using"         GAMMA CAT. ##.###";Gamma3
1187    Lprint Tab(23):Lprint Using"GAMMA OH  ##.###";Gammaoh;
1188    Lprint Using"         GAMMA AN.  ##.###";Gamma4
1190    If Concads=0 Then Goto 1262
1200    Suma=0 :   For M=1 To 3
1210      Adsion(M)=Maxada*Ads(M)/(1+Sumads)
1220      Suma=Suma+Adsion(M)
1230      Next M

1240    Print:Lprint
1245    Print Using" TOTAL ADSORPTION (MICRO M/SQ.M.)        ##.###";Suma
1246    Lprint Using" TOTAL ADSORPTION (MICRO M/SQ.M.)        ##.###";Suma
1247    Lprint:Lprint
1248    Print : Print
1249    Print Tab(21);"MONOVALENT";Tab(38);"DIVALENT";Tab(52);"TRIVALENT"
1250    Print" PROPN. ADSORBED";
1251    Print Using"        ###.####";Adsion(1)/Suma,Adsion(2)/Suma,
1252    Print Using"        ###.####";Adsion(3)/Suma
1254    Print" AMOUNT ADSORBED";
1256    Print Using"        ##.####";Adsion(1),Adsion(2),Adsion(3)
1257    Print" PROPN. IN SOLN.";
1258    Print Using"        ###.####";Concion(1)/Concads,Concion(2)/Concads,
1259    Print Using"        ###.####";Concion(3)/Concads



1260    Print  :Print Using"NO OF OUTER CYCLES   ##";Ii
1262    Print Using"NO OF EXTRAPOLATIONS ##";I : Print : Print


1270    Print"FOR NEW VALUES OF: CHARGE PARAMETERS INPUT 1"
1275    Print"                 ADS. PARAMETERS    INPUT 2"
1280    Print"                 BACKGROUND ELECT.  INPUT 3"
1295    Print"                 CONCENT. OR pH     INPUT 4"
1297    Print" TO QUIT INPUT 5"


1300    Input Index
1310    On Index Goto 270,350,410,440,1400
1320    If Index=0 Or Index>5 Goto 1270


1400    End

2000    Print "ERROR NO",Err," AT LINE NO ",Erl:Resume Next
```

117

Chapter B3

Deriving equations to describe adsorption and rate of adsorption

In modelling the adsorption of ions by charged surfaces, four distinct situations are of potential interest. They are:

1) equilibrium adsorption with the equilibrium concentration of adsorbate specified;
2) equilibrium adsorption with the initial concentration of adsorbate and adsorbant specified;
3) rate of adsorption with the concentration of adsorbate held constant;
4) rate of adsorption with the initial concentration of adsorbate and adsorbant specified.

Models utilizing all of these situations will be presented (in Chapters B2, B4, B6 and B4 respectively). However it is convenient to present here a common development of the four relevant equations.

Common development

Let us assume that a solution containing an adsorbate is brought into contact with a solid adsorbent. For convenience we will assume that only one of the ionic species present in solution reacts with the surface. If more than one species should react, an additive process will give the appropriate equations. The overall reaction involved may comprise a number of steps — for example the reaction of the ion with the surface, the displacement of a water molecule from the surface, gain or loss of protons, and/or the approach or departure of an electrolyte ion to balance the change in charge of the surface. We do not know, from first principles, what the sequence of these reactions will be. However we assume that one of the reaction steps is appreciably slower than the others and that the rate of this reaction therefore determines the overall rate. For a multistep reaction at a charged surface the rate equation (obtained from Bockris and Reddy 1970, equation 9.24) is:

$$dc/dt = - k_1 \alpha \gamma c_t \, m_t \, \exp(-sgn(z) \, \overleftarrow{\alpha} \, F\psi_a/RT) + k_2 \, s_t \, \exp(sgn(z) \, \overrightarrow{\alpha} \, F\psi_a/RT)$$

(B3.1)

where the symbols are defined in Table B3.1 and

$$\overleftarrow{\alpha} = (n - \overrightarrow{\gamma})/\nu - r_\beta$$

(B3.2)

and

$$\overrightarrow{\alpha} = \overrightarrow{\gamma}/\nu + r_\beta$$

(B3.3)

Further

$$m_t = (N_T - \Gamma_a) \, AW$$

(B3.4)

and

118

Table B3.1

Tabulation of the symbols used in the equations of Chapter B3

A	surface area of the adsorbent
W	mass of adsorbent (g) per litre of solution
N_T	the surface density of adsorption sites (mol/m^2)
Γ_a	the surface density of adsorbate (mol/m^2)
c	the concentration of total adsorbate in solution (mol/l)
m	the concentration of vacant sites (mol/l)
s	the concentration of occupied sites (mol/l)

o	subscripts used to indicate concentrations at time zero,
e	at equilibrium, at time t, and at the beginning of a step,
t	respectively
i	

α	proportion of the adsorbate present as the ionic species that reacts with the surface
γ	activity coefficient for the adsorbing ion in solution
z	the valency of the adsorbing ion
$\text{sgn}(z)$	the sign of the valency of the adsorbing ion (i.e. plus or minus)
ψ_a	electric potential in the plane of adsorption (mV)
R	the gas constant
F	the Faraday
T	the temperature (K)
k_1	the rate constant for the forward reaction $(\text{min}^{-1}\ \text{litre mol}^{-1})$
k_2	the rate constant for the back reaction (min^{-1})
$\overleftrightarrow{\alpha}$	transfer coefficient for the forward reaction at a charged surface (Bockris and Reddy 1970, eq. 9.25)
$\overrightarrow{\alpha}$	transfer coefficient for the back reaction at a charged surface
n	the total number of electrons involved in the overall reaction
ν	the number of times the rate determining step must occur for the overall reaction to occur once
r	the number of electrons transferring in the rate determining step
$\overrightarrow{\gamma}$	the number of single electron transfer steps that preceded the rate determining step
β	is the symmetry factor. This factor partitions the work done on a charged particle into that due to the approach to the surface and that due to the departure from the surface. Further descriptions of the coefficients listed between $\overleftrightarrow{\alpha}$ and β are given in Bockris and Reddy (1970).

$$s_t = \Gamma_a \, AW \tag{B3.5}$$

Equilibrium adsorption with equilibrium concentration specified

At equilibrium the rate of reaction is zero and hence, from the equations B3.1, B3.4 and' B3.5:

$$\frac{\Gamma_a}{N_T - \Gamma_a} = \frac{k_1}{k_2} \, \alpha \, \gamma \, c_e \exp(\overleftarrow{\alpha} + \overrightarrow{\alpha})(\text{-sgn}(z)\,F\psi_a/RT)) \tag{B3.6}$$

From B3.2 and B3.3, $\overleftarrow{\alpha} + \overrightarrow{\alpha} = n/\nu$ — that is, the total number of electrons involved divided by the number of times the rate determining step must occur. It is reasonable to assume that the number of electrons is equal to the charge in the ion (that is $n = z$) and that $\nu = 1$. Further, the equilibrium coefficient, K, is equal to the ratio k_1/k_2. Hence

$$\frac{\Gamma_a}{N_T - \Gamma_a} = K \, \alpha \, \gamma \, c_e \exp(-z\,F\psi_a/RT) \tag{B3.7}$$

This expression was derived by Bowden *et al.* (1977) using a different argument. From it, the equation relating adsorption to the equilibrium concentration can be obtained:

$$\Gamma_a = \frac{N_T \, \alpha \, \gamma \, c_e \exp(-z\,F\psi/RT)}{1 + \alpha \, \gamma \, c_e \exp(-z\,F\psi/RT)} \tag{B3.8}$$

This is the first of the four required equations and is used in Chapter B2.

Equilibrium adsorption with the initial concentration specified

For convenience, let

$$K^* = K \, \alpha \, \gamma \exp(-z\,F\psi_a/RT) \tag{B3.9}$$

Then using B3.4 and B3.5, equation B3.7 can be written

$$s_e/m_e = c_e \, K^* \tag{B3.10}$$

By mass balance:

$$m_\sigma - m_o = c_e - c_o = s_o - s_e \tag{B3.11}$$

Hence substituting for s_e and for m_e using B3.11 gives:

$$(s_o - c_e + c_o)/(c_e - c_o + m_o) = c_e \, K^* \tag{B3.12}$$

Rearranging gives a quadratic equation in c_e:

$$c_e^2 \, K* + c_e \, (m_o \, K* - c_o \, K* + 1) - (s_o + c_o) = 0 \qquad (B3.13)$$

The positive root of this equation relates the equilibrium concentration of adsorbate (c_e) to the initial concentrations of adsorbate (c_o), of vacant sites (m_o), and of occupied sites (s_o). From the change in concentration, the amount of adsorption can be calculated. This equation can therefore be used to relate adsorption to the initial conditions (Barrow *et al.* 1981a). It is used in the program called "ZNCL2".

Rate of adsorption with concentration of adsorbate constant

Again, for convenience, let

$$k_1^* = k_1 \, \alpha \, \gamma \, \exp\left((-\text{sgn}(z)) \, \overline{\alpha} \, F\psi / RT\right) \qquad (B3.14)$$

and

$$k_2^* = k_2 \, \exp\left(((-\text{sgn}(z)) \, \overrightarrow{\alpha} \, F\psi_a / RT\right) \qquad (B3.15)$$

Then equation (B3.1) may be written:

$$ds/dt = k_1^* \, c_t \, m_t - k_2^* \, s_t \qquad (B3.16)$$

If the concentration c_t is held constant, this equation describes opposing first order reactions. The integration of such equations is given in most text books on physical chemistry. However, in this case, integration is not straightforward because reaction between an ion and a charged surface will change the electric potential of the surface. Nevertheless the reaction can be followed through time using a series of steps each of which is small enough for the potential (ψ_a) to be regarded as constant. During each of these steps (Δ_t) the change in adsorption (Δ_s) will be given by:

$$\Delta_s = \frac{k_1^* \, c \, m_i - k_2^* \, s_i}{k_1^* \, c + k_2^*} \left[1 - \exp\left(-\Delta_t \, (k_1^* \, c + k_2^*)\right) \right] \qquad (B3.17)$$

This equation relates the change in adsorption during a time step to the initial concentration of vacant (m_i) and occupied sites (s_i) when the solution concentration of adsorbate is constant. It is used in the program called "DIFPLUS" (Chapter B6).

Rate of adsorption with initial concentration specified

Finally let us rewrite equation B3.16 as:

$$dc/dt = -k_1^* \, c_t \, m_t + k_2^* \, s_t \qquad (B3.18)$$

Substituting for m_t and s_t using B3.11 and rearranging gives:

$$dc/dt = -k_1^* c_t^2 + (k_1^* c_o - k_1^* m_o - k_2^*) + k_2^* (c_o + s_o) \qquad (B3.19)$$

Let $\quad a = c_o - m_o - k_2^* / k_1^* \qquad\qquad\qquad\qquad\qquad\qquad\qquad (B3.20)$

and

$$b = (c_o + s_o) k_2^* / k_1^* \qquad\qquad\qquad\qquad\qquad\qquad (B3.21)$$

Then equation B3.19 becomes

$$dc/dt = k_1^* (-c_t^2 + a c_t + b) \qquad\qquad\qquad\qquad (B3.22)$$

Integration of this equation poses similar problems to that of B3.16 but again it can be integrated over intervals of time that are small enough for the electric potential (ψ_a) to be regarded as constant. Integration over the interval $t = 0$ to $t = t$ then gives:

$$c_t = (q + B c_e)/(B - 1) \qquad\qquad\qquad\qquad (B3.23)$$

where

$$B = \exp\left[(c_e + q) k_1^* (t - t_o)\right] (c_o + q)/(c_o - c_e)$$

$$q = [- a + \sqrt{(a^2 + 4b)}]/2$$

$$c_e = [a + \sqrt{(a^2 + 4b)}]/2$$

Equation B3.23 gives the change in concentration after a given time interval. From this change, the amount of adsorption can be calculated and thus the starting condition for the next time interval. It was used for this purpose by Barrow et al. (1981b). It is used in the program called "RATEOX".

Values for $\overline{\alpha}$ and $\overrightarrow{\alpha}$

Finally, let us consider the values that might be assigned to $\overline{\alpha}$ and $\overrightarrow{\alpha}$. These terms are components of k_1^* and of k_2^* and so are involved in specifying the rates of the reactions. The values to be assigned to $\overline{\alpha}$ and $\overrightarrow{\alpha}$ depend on the sequence of the steps involved in the overall reaction. It was shown, for phosphate reaction with goethite, that appropriate values were $\overline{\alpha} = 2$ and $\overrightarrow{\alpha} = 0$ (Barrow et al. 1981b). This was taken to mean that the rate-determining step preceded the electron transfer step and did not, of itself, involve electron transfer. No appropriate experiments have been published for other adsorbates and hence when a value is needed for a modelling study, it has been assumed that the behaviour is analogous to that of phosphate.

References

Barrow, N.J., Madrid, L. and Posner, A.M. 1981a. A partial model for the rate of adsorption and desorption of phosphate by goethite. Journal of Soil Science 32, 399-407.

Barrow, N.J., Bowden, J.W., Posner, A.M. and Quirk, J.P. 1981b. Describing the adsorption of copper, zinc and lead on a variable charge mineral surface. Australian Journal of Soil Research 19, 309-321.

Bockris, J. O'M. and Reddy, A.K.N. 1970. Modern Electrochemistry. Plenum Press, New York.

Bowden, J.W., Posner, A.M. and Quirk, J.P. 1977. Ionic adsorption on variable charge mineral surfaces. Theoretical charge development and titration curves. Australian Journal of Soil Research 15, 121-136.

Chapter B4

Modifications to the four layer model

Two variations of the model BOWDEN described in Chapter B2 are presented here. The original model requires, as an input, the equilibrium concentration of adsorbate. This format is useful for studying the effects of varying the values of the parameters and of varying the conditions used. It can also be useful for comparing the model with data — several publications report adsorption in terms of equilibrium concentration of adsorbate. However it is also convenient to be able to relate adsorption to the initial conditions. This is the format of the present version. This format is also useful for comparing the model with data as there are also many publications in which adsorption is reported in terms of the initial concentration. This version of the model, ZNCL2, is supplied in a form suitable for modelling the reaction of Me^{2+} $MeOH^+$, and/or, $MeCl^+$ ions with a surface. The reasons for this choice are firstly that this is a common situation, and secondly that it illustrates how to deal with cases in which complexes such as $MeCl^+$ may also be involved in the reaction. The algebraic development of the equations used to relate adsorption to the initial concentration is given in Chapter B3.

The other (RATEOX) variation considered permits the rate of the reaction with oxides to be modelled. The rate is assumed to have two components. The first is the rate of the initial adsorption reaction. This is based on the algebraic development of Chapter B3. Adsorption is then assumed to initiate a diffusive movement into the particle as discussed in Chapter A4. For phosphate, the data treated in that Chapter suggest that the rate of movement is slow — and accordingly a low value is suggested for the diffusion term. A higher value may be appropriate for other ions. For use with metal ions, the program should be changed — as indicated in "ZNCL2". As for that program, the model requires the original concentration of the reactants as an input.

The program, as written, requires the operator to enter the time steps. At any stage the operator may decide to change the conditions as, for example, to increase the dilution and so induce desorption. For routine operation it may be convenient to enter a sequence of times as a "Data" statement and to replace the "Input" statements by "Read" statements.

```
1    Rem                ZNCL2.BAS



10    Print "THIS PROGRAM IS DERIVED FROM THE BOWDEN MODEL FOR ADSORPTION."
15    Lprint"THIS PROGRAM IS DERIVED FROM THE BOWDEN MODEL FOR ADSORPTION."
16    Print "IT IS FOR METAL IONS WITH INITIAL CONC SPECIFIED."
17    Lprint "IT IS FOR METAL IONS WITH INITIAL CONC SPECIFIED."
19    Lprint:Lprint
20    Print "THERE IS PROVISION FOR NON-SYMETRICAL ELECTROLYTES & IT CAN BE"
30    Print"USED EITHER IN THE PRESENCE OR THE ABSENCE OF SPECIFIC ADSORPTION."
45    Print "SOME SUGGESTED (ROUNDED) VALUES OF THE PARAMETERS ARE GIVEN "



50    Rem FOR SIMPLER MODELS SET Kcat=0, Kan=0 & / OR SET APPROPRIATE
55    Rem VALUES OF CAPACITANCE (Gsa,Gac,Gcd) LARGE - E.G. 1e10
60    Rem SYMBOLS FOR CHARGE, POTENTIAL, & CAPACITANCE ARE DEFINED IN OUTPUT.
70    Rem Kh IS K(H), Koh IS K(OH), Kcat IS K(Cation), Kan IS K(Anion),
90    Rem Kdiss1,Kdiss2 ARE DISSOCIATION CONSTS OF THE METAL.
100   Rem Kass IS ASSOC CONST FOR METAL ANION COMPEX eg METAL-CHLORIDE.
110   Rem Kaff IS AN ARRAY OF AFFINITY COEFF FOR THE DIVALENT,MONOVALENT,
115   Rem UNCHARGED & MeCl IONS
120   Rem Ads & Chg ARE ARRAYS OF TERMS IN THE BOWDEN ADSORPTION CHARGING EQN.
130   Rem Adsion IS AN ARRAY OF ADSORPTION OF INDIVIDUAL ADSORBED SPECIES



140   Dim Act(3),Concion(3),Kaff(3),Chg(3),Ads(3),Adsion(3),Kstar(4)
150   On Error Goto 2000



160   Rem THE NEXT STATEMENTS CONTROL INPUTS OF PARAMETERS



170   Print" MODEL IS REPRESENTED:"
180   Print" SURFACE        ADSORBED        COUNTER         DIFFUSE"
190   Print"  PSI(S)         PSI(I)          PSI(C)          PSI(D)"
200   Print"  Psi1           Psi2            Psi3            Psi4"
210      For I=1 To 5
212      Print"     |             |               |              |"
214      Next I
220   Print"       | CAPTNCE Gsa | CAPTNCE Gac|  CAPTNCE Gcd |"
230   Print" SIGMA(S)        SIGMA(I)        GAMMA(NA)       SIGMA(D)"
240   Print"                                 GAMMA(CL)"
250   Print"  Sigma1          Sigma2         Gamma3 & 4      Sigma4
```

```
260    Print : Print : Print : Print"VALUES OF PARAMETERS"
265    Index=0
270    Print"MAX ADSORPTION IN S PLANE IN MICROMOLES/SQ.M ( 10 )   ",Tab(55)
271    Input Maxadss
272    Lprint"MAX ADSORPTION IN S PLANE IN MICROMOLES/SQ.M",Tab(55), Maxadss
275    Print "CAPACITANCES TO BE ENTERED AS FARADS/SQ.M."
280    Print"CAPACITANCE Gsa( 3 )",Tab(55); : Input Gsa
285    Lprint"CAPACITANCE Gsa ",Tab(55), Gsa
290    Print"CAPACITANCE Gac( 5 )",Tab(55); : Input Gac
295    Lprint"CAPACITANCE Gac",Tab(55),Gac
300    Print"CAPACITANCE Gcd(V. LARGE eg 1e12 or 0.2 )",Tab(55); : Input Gcd
305    Lprint"CAPACITANCE Gcd",Tab(55),Gcd
310    Print"CHECK - CAPACTNCE FROM SURFACE TO COUNTER IONS IS",1/(1/Gsa+1/Gac)


315    Print"IT IS ASSUMED THAT THE DISSOCIATION CONSTANTS HAVE BEEN "
316    Print" OBTAINED FROM BAES & MESMER(SEE REFS CHAPTER A1)"
317    Print "& THEREFORE WILL HAVE BEEN CORRECTED FOR IONIC STRENGTH"


319    Print:Print
330    Print"K(H) ( 1e7 )",Tab(55):Input Kh
331    Print"K(OH) ( 1e4 )",Tab(55):Input Koh
333    Lprint"K(H) ",Tab(55),Kh : Lprint"K(OH)",Tab(55),Koh
335    Print"K(CAT) ( 1 )",Tab(55);
336    Input Kcat : Print"K(AN) ( 1 )",Tab(55); : Input Kan
337    Lprint"K(CAT)",Tab(55),Kcat : Lprint"K(AN)",Tab(55),Kan
338    Lprint:Lprint


340    Input "ARE SPECIFICALLY ADSORBING IONS PRESENT (Y/N)?",Adsrp$
342     If Adsrp$="N" Or Adsrp$="n" Goto 410
 345    If Index>0 goto 440
350    Print"MAX. ADS IN PLANE A IN MICROMOLES/SQ M.(FOR Zn TRY 6 )",Tab(55);
353    Input Maxada
355    Lprint"MAX. ADS IN PLANE A IN MICROMOLES PER SQ M. ",Tab(55),Maxada
385    Print"DISS CONSTS FOR METALS PLUS ASSOC. CONST FOR ANION COMPLEX "
386    Print"SUCH AS MeCl. NOTE  CONSTANTS - NOT LOG CONST"
387    Print "INDICATED VALUES ARE FOR Zn IN 0.01M NaCl"
390    Lprint"DISS & ASSOC CONSTS FOR ADSORBING SPECIES. "
391    Print"FIRST DISS CONST ( 9.03E-10) ",Tab(55); : Input Kdiss1
392    Lprint"FIRST DISS CONST",Tab(55),Kdiss1
393    Print"SECOND DISS CONST (1.15E-8 )",Tab(55); : Input Kdiss2
394    Lprint"SECOND DISS CONST",Tab(55),Kdiss2
395    Print"ASSOCIATION CONST (0.8 )",Tab(55); : Input Kass
396    Lprint"ASSOCIATION CONST",Tab(55), Kass
```

```
397    Print"BINDING CONSTANTS FOR ADSORBING SPECIES. ENTER ZEROES FOR NON "
398    Print"ADSORBED SPECIES. USE MICROMOLAR SCALE ie L/MICROMOLE"
399    Lprint:Lprint"BINDING CONSTANTS FOR ADSORBING SPECIES.(L/MICROMOLE)"
400    Print"SUGGESTED VALUES ARE FOR Zn"
401    Print"Me++ (0 )";Tab(55); : Input Kaff(0)
402    Lprint"Me++ ";Tab(55), Kaff(0)
405    Print"Me(OH)+ ( 3) )";Tab(55); : Input Kaff(1)
406    Lprint"Me(OH)+";Tab(55),Kaff(1)
407    Print"Me(OH)2 ( 0 )",Tab(55); : Input Kaff(2)
408    Lprint"Me(OH)2";Tab(55), Kaff(2)
409    Print "MeCl+ ( 1 )",Tab(55);:Input Kaff(3)
410    Lprint "MeCl+ ";Tab(55),Kaff(3)


412    Print "HYDROLYSIS CONSTANT FOR SOLID Me(OH)2 (FOR Zn C 3e12 )",Tab(55);
413    Input Qsmeoh
414    Lprint "HYDROLYSIS CONSTANT FOR SOLID Me(OH)2 ",Tab(55),Qsmeoh
415    Rem SUGGEST USE CONSTANTS FOR AMORPHOUS FORMS
417    If Index>0 Goto 440


418    Print"VALENCY OF ELECTROLYTE CATION",Tab(55); : Input Vcat
419    Lprint:Lprint"VALENCY OF ELECTROLYTE CATION",Tab(55),Vcat
420    Print"VALENCY OF ELECTROLYTE ANION",Tab(55); : Input Van
421    Lprint"VALENCY OF ELECTROLYTE ANION",Tab(55),Van


422    Print : Print"  VALUES OF VARIABLES"
423    Lprint:Lprint
424    Print "WT. OF VARIABLE CHARGE OXIDE IN G/LITRE ( SAY 0.5 )",Tab(55);
425    Input Wt
426    Lprint "WT. OF V.C. OXIDE IN G PER LITRE" ,Tab(55),Wt
427    Print "SURFACE AREA OF ADSORBATE IN SQ. M/G ( SAY 80 )",Tab(55);
428    Input Area
429    Lprint "SURFACE AREA OF ADSORBATE IN SQ. M/G ",Tab(55), Area
430    Concsites=Maxada*Wt*Area
440    Print"pH",Tab(55); : Input Ph
442    Lprint"pH",Tab(55),Ph
445    Print"ELECTROLYTE CONC.(MOLAR SCALE)",Tab(55); : Input Concelec
446    Lprint"ELECTROLYTE CONC.(MOLAR SCALE)",Tab(55), Concelec


450    Rem SIMPLE CALCULATION OF ACTIVITY COEFF FOLLOWS.
455    Rem THIS SIMPLE VERSION WILL NOT GIVE GOOD VALUES FOR CONC ELECTROLYTES
456    Rem SEE 475 FOR A WAY TO DEAL WITH THIS


460    Mu=0.5*(Van*Concelec*Vcat^2+Vcat*Concelec*Van^2)
```

```
470   Actcoef=10^(-0.5*Sqr(Mu)/(1.0+Sqr(Mu)))
475   If Concelec=1 Then Actcoef=0.658
476   Rem  A SIMPLE WAY TO DEAL WITH CONC ELECTROLYTES
500   Kdiss3=1/(Kass*Actcoef^4)
505   If Adsrp$="Y" Or Adsrp$="y" Then 510 Else 512
510   Print"INITIAL CONC. OF METAL IN MICROMOLES/L";Tab(55); : Input Concads
511   Lprint"INITIAL CONC. OF METAL IN MICROMOLES/L";Tab(55),Concads
512   Print : Print


520   Rem THE NEXT STATEMENTS MAKE A FIRST ESTIMATE OF PSI(S).
530   ConcH=10^(-Ph) : ConcOH=1E-14/ConcH
550   Term=Sqr(Koh*1E-14/Kh) : Zpc=-Log(Term)/Log(10)
560   Psil=58*(Zpc-Ph)
570   Print Using"NERNST EST  ####.#";Psil


580   Rem THE NEXT STATEMENTS ARE THE SET OF CHARGING EQUATIONS
590   Rem THEY END AT 980 WITH A SECOND ESTIMATE OF SIGMA)(S) - Estsigma1
600   Print
610   Delta1=10:Ii=0
620   I=1
630   Hterm=Kh*ConcH*Exp(-0.039*Psil)
640   Ohterm=Koh*ConcOH*Exp(0.039*Psil)
641   Gammah=Maxadss*Hterm/(1+Hterm+Ohterm)
642   Gammaoh=Maxadss*Ohterm/(1+Hterm+Ohterm)
650   Sigma1=Gammah-Gammaoh
660   Psi2=Psil-Sigma1*96.487/Gsa
661   Rem USE FARADAY CONSTANT TO GIVE POTNL IN MILLIVOLTS
670   If Concads=0 Then Sigma2=0 : Goto 770
675   Meclterm=ConcH^2*Concelec*Vcat
680   T1=ConcH^2*Kdiss3 +ConcH*Kdiss1*Kdiss3 +Kdiss1*Kdiss2*Kdiss3 +Meclterm
681   Concion(0)=ConcH^2*Kdiss3/T1
682   Concion(1)=ConcH*Kdiss1*Kdiss3/T1
683   Concion(2)=Kdiss1*Kdiss2*Kdiss3/T1
684   Concion(3)=ConcH^2*Concelec*Vcat/T1
689   Alphagamma(0)=Actcoef^4*Concion(0)
690   Alphagamma(1)=Actcoef*Concion(1)
710   Alphagamma(2)=Concion(2)
730   Alphagamma(3)=Actcoef*Concion(3)


745   Rem  THE NEXT SECTION CALCULATES THE CHARGE DUE TO ASORPTION
750   Sigma2=0 : Sumkstar=0
752    For K=0 To 3
753      Z=2-K:If K=3 Then Z=1
755      Kstar(K)=Kaff(K)*Alphagamma(K)*Exp(-Z*0.039*Psi2)
756      Sumkstar=Sumkstar+Kstar(K)
```

```
757     Next K
758     B=Concads-Concsites-1.0/Sumkstar
759     C=Concads/Sumkstar
760     Q=(B+Sqr(B^2+4*C))/2
761     Suma=(Concads-Q)/(Wt*Area)
762     For K=0 to 3
763      Z=2-K:If K=3 Then Z=1
764      Adsion(K)=Suma*Kstar(K)/Sumkstar
765      Sigma2=Sigma2+Z*Adsion(K)
766     Next K
770     Psi3=Psi2-(Sigma1+Sigma2)*96.487/Gac


800     Rem   CALC MOLARITY OF INDIVIDUAL IONS FOR NON SYMMETRICAL ELECTROLYTE.
810     Conccat=Concelec : If Van>Vcat Then Conccat=Concelec*Van
820     Conccat=Conccat*Actcoef^(Vcat^2)
840     Concan=Concelec : If Vcat>Van Then Concan=Concelec*Vcat
850     Concan=Concan*Actcoef^(Van^2)
860     Anionterm=Kan*Concan*Exp(0.039*Van*Psi3)
865     Cationterm=Kcat*Conccat*Exp(-0.039*Vcat*Psi3)
880     Gamma4=Maxadss*Van*Anionterm/(1+Anionterm+Cationterm)
890     Gamma3=Maxadss*Vcat*Cationterm/(1+Anionterm+Cationterm)
900     Psi4=Psi3-(Sigma1+Sigma2+Gamma3-Gamma4)*96.487/Gcd
930     If Vcat=Van Then 931 Else 940
931     T5=Vcat*0.0195*Psi4
932     If Abs(T5)>87 Then Sigma4=-Sgn(Psi4)*1e35: Goto 980
934     T6=0.5*(Exp(T5)-Exp(-T5))
936     Sigma4=-1.22*Sqr(Concelec)*T6
938     Goto 980
940     X1=0.039*Psi4*Vcat : X2=0.039*Psi4*Van
945     If Abs(X1)>87 Then Sigma4=-Sgn(Psi4)*1e35: Goto 980
950     W3=(Exp(-X1)-1)/Vcat : W4=(Exp(X2)-1)/Van
960     W5=Sqr(W3+W4) : If Psi4>0 Then W5=-W5
970     Sigma4=.61*Sqr(Concelec)*W5
980     Estsigma1=-Sigma2-Gamma3+Gamma4-Sigma4
985     Rem   END OF CALCULATION OF CHARGING EQUATIONS


990     Rem   NOW COMPARE THE 2 ESTIMATES OF SIGMA(S) & ADJUST PSI(S)
1000    Difference=Sigma1-Estsigma1
1005    If Abs(Difference)<1E-04 Then Goto 1134
1006    Delta2=Sgn(Difference)*Delta1


1007    Rem IF DIFFERENCE LARGE USE INCREMENTAL SEARCH
1008    If Abs(Difference)>100 Goto 1110
```

```
1009    Rem WHEN DIFFERENCE SMALL ENOUGH USE LINEAR EXTRAPOLATION
1010    If I=1 Then Prepsi=Psi1 : Goto 1090
1020    Slope=(Difference−prediff)/(Prepsi−Psi1)
1030    Intercept=Difference+Slope*Psi1
1035    Prepsi=Psi1
1037    If Slope=0 then Psi1=Prepsi*0.999: Goto 1090
1040    Psi1=Intercept/Slope
1050    If Psi1=Prepsi Then Psi1=Psi1*0.999
1090    Prediff=Difference
1091    If I=11 Then 1092 Else 1094
1092    Print "SLOW CONVERGENCE. PRINT OUT SUCCESSIVE ESTS OF POTNL & DIFFERENCE"
1094    If I>10 Then Print I,Prepsi,Difference
1095    If I>20 And Abs(Difference)<1e−3 Goto 1134
1096    If I=30 Then 1097 Else 1100
1097    Print "EQUATIONS WON'T CONVERGE.  MAYBE THE CAPACITANCE Gcd IS TOO "
1098    Print "SMALL GIVING HIGH SENSITIVITY TO PSI":Goto 1270
1100    I=I+1
1105    If I=2 Then Psi1=Psi1+Delta2
1106    Goto 630
1108    Rem THIS IS THE END OF THE EXTRAPOLATION


1109    Rem THIS IS THE BEGINNING OF THE ALTERNATIVE INCREMENTAL SEARCH
1110    Ii=Ii+1
1115    If Ii=1 Then 1128
1125    Prod=Sgn(Difference)*Sgn(Prediff)
1126    If Prod<0 Then Delta1=0.49*Delta1
1128    Prepsi=Psi1:Psi1=Psi1+Delta2
1129    prediff=Difference
1130    Goto 630
1132    Rem END OF INCREMENTAL SEARCH


1134    Rem THE NEXT STATEMENTS CONTROL OUTPUT.
1136    Lprint:Lprint
1137    Residsol=Concads−Suma*Wt*Area
1138    Me2plus=Residsol*Alphagamma(0)
1139 .  If (Me2plus*1e−6/ConcH^2)<Qsmeoh Then 1150
1140    Print "PRECIPITATION OCCURRED":Lprint "PRECIPITATION OCCURRED"
1141    Solnconc=Qsmeoh*ConcH^2*1e6/Alphagamma(0)
1142    Print "METAL CONC IN SOLN. (MICROMOLES/L) ",Solnconc
1143    Lprint"METAL CONC IN SOLN. (MICROMOLES/L) ",Solnconc
1144    Apparents=(Concads−Solnconc)/(Wt*Area)
1145    Print "METAL APPARENTLY ADSORBED (MICROMOLES/SQ.M.)  ",Apparents
1146    Lprint "METAL APPARENTLY ADSORBED (MICROMOLES/SQ.M.) ",Apparents
1147    Print "TO FIND HOW MUCH WAS REALLY ADSORBED RUN A MODEL WITH THE"
1148    Print" INDICATED SOLUTION CONC ENTERED AS FINAL CONC"
1149    Goto  1270
```

```
1150   Print Tab(31)"SURFACE        ADSORBED        COUNTER        DIFFUSE"
1160   Print"POTENTIAL (M.VOLTS)        ";
1162   Print Using"      ####.###";Psi1,Psi2,Psi3,Psi4
1164   Print"CHARGE (MICRO M/SQ.M.)    ";
1168   Print Using"        ##.###";Sigma1,Sigma2,Gamma3-Gamma4,Sigma4
1170   Print Tab(23):Print Using"GAMMA H    ##.###";Gammah;
1172   Print Using"        GAMMA CAT. ##.###";Gamma3
1174   Print Tab(23):Print Using"GAMMA OH  ##.###";Gammaoh;
1176   Print Using"        GAMMA AN.  ##.###";Gamma4
1178   Lprint Tab(31)"SURFACE        ADSORBED        COUNTER        DIFFUSE"
1180   Lprint"POTENTIAL (M.VOLTS)        ";
1182   Lprint Using"      ####.###";Psi1,Psi2,Psi3,Psi4
1184   Lprint"CHARGE (MICRO M/SQ.M.)    ";
1186   Lprint Using"       '##.###";Sigma1,Sigma2,Gamma3-Gamma4,Sigma4
1188   Lprint Tab(23):Lprint Using"GAMMA H    ##.###";Gammah;
1190   Lprint Using"        GAMMA CAT. ##.###";Gamma3
1192   Lprint Tab(23):Lprint Using"GAMMA OH  ##.###";Gammaoh;
1194   Lprint Using"        GAMMA AN.  ##.###";Gamma4
1196   If Concads=0 Then Goto 1260


1200   Print:Lprint
1205   Print Using"TOTAL ADSORPTION (MICRO M/SQ.M.)        ##.###";Suma
1210   Lprint Using" TOTAL ADSORPTION (MICRO M/SQ.M.)        ##.###";Suma
1215   Print Using"NET SURFACE CHARGE                      ##.###";Sigma1+Sigma2
1220   Lprint Using" NET SURFACE CHARGE                      ##.###";Sigma1+Sigma2

1225   Percntads=(Concads-Residsol)*100/Concads
1230   Print Using"PERCENT OF ADDED METAL ADSORBED        ###.###";Percntads
1235   Lprint Using" PERCENT OF ADDED METAL ADSORBED        ###.###";Percntads
1240   Print Using"METAL REMAINING IN SOLUTION          #####.###";Residsol
1245   Lprint Using" METAL REMAINING IN SOLUTION          #####.###";Residsol
1250   Lprint:Lprint
1251   Print Tab(24);"Me++";Tab(37);"Me(OH)+";Tab(50);"Me(OH)2";Tab(63);"MeCl+"


1252   Print" PROPN. ADSORBED";
1253   Print Using"      ###.####";Adsion(0)/Suma,Adsion(1)/Suma;
1254   Print Using"      ###.####";Adsion(2)/Suma,Adsion(3)/Suma
1255   Print" AMOUNT ADSORBED";
1257   Print Using"        ##.####";Adsion(0),Adsion(1),Adsion(2),Adsion(3)
1258   Print" PROPN. IN SOLN.";
1259   Print Using"      ###.####";Concion(0),Concion(1),Concion(2),Concion(3)
1260   Print  :Print Using"NO OF OUTER CYCLES   ##";Ii
1262   Print Using"NO OF EXTRAPOLATIONS ##";I : Print


1270    Print"FOR NEW VALUES OF: CHARGE PARAMETERS INPUT 1"
```

```
1275    Print"                    ADS. PARAMETERS    INPUT 2"
1280    Print"                    BACKGROUND ELECT.  INPUT 3"
1295    Print"                    CONCENT. OR pH     INPUT 4"
1297    Print" TO QUIT INPUT 5"
1300    Input Index


1310    On Index Goto 270,350,418,440,1400
1320    If Index=0 Or Index>5 Goto 1270
1400    End

2000    Print  "ERROR NO ",Err,"  AT LINE NO  ",Erl: Resume Next
```

```
1    Rem              RATEOX.BAS


10   Print "THIS PROGRAM IS THE EXTENDED BOWDEN MODEL FOR ADSORPTION."
15   Print "TO WHICH THE EQUATIONS FOR RATE OF ADS & RATE OF PENETRATION "
20   Print "HAVE BEEN ADDED. IT IS BASED ON CONSANT AMOUNT OF ADSORBATE "
21   Print "RATHER THAN CONSTANT CONC."
23   Print "SOME SUGGESTED VALUES OF THE PARAMETERS ARE PROVIDED FOR P"
25   Lprint "THIS PROGRAM IS THE EXTENDED BOWDEN MODEL FOR ADSORPTION."
30   Lprint "TO WHICH THE EQUATIONS FOR RATE OF ADS & RATE OF PENETRATION "
35   Lprint "HAVE BEEN ADDED. IT IS BASED ON CONSANT AMOUNT OF ADSORBATE "
36   Lprint "RATHER THAN CONSTANT CONC."
40   Print "THERMODYNAMIC DISSOCIATION CONSTANTS FOR P ARE SUPPLIED."
45   Print "TO WORK WITH OTHER NUTRIENTS, THE CONSTANTS MUST BE SUPPLIED."
50   Input"PRESS RETURN WHEN READY TO PROCEED",Qqq


60   Rem SYMBOLS FOR CHARGE, POTENTIAL, & CAPACITANCE ARE DEFINED IN OUTPUT.
70   Rem Kh IS K(H), Koh IS K(OH), Kcat IS K(Cation), Kan IS K(Anion),
90   Rem Kdiss1,Kdiss2,& Kdiss3 ARE DISSOCIATION CONST OF PHOSPHORIC ACID
100  Rem OR OF OTHER ADSORBATE AS ENTERED
110  Rem Kaff IS AN ARRAY OF AFFINITY COEFF FOR MONO, DI, & TRIVALENT IONS
120  Rem M IS AN ARRAY OF CONCS OF THESE IONS
130  Rem Adsion IS AN ARRAY OF ADSORPTION OF INDIVIDUAL ADSORBED SPECIES


140  Dim Act(3),Concion(3),Kaff(3),Kdiss(3)
142  Dim Occsites(50),Coefdiff(50),Time(50),Sumprod(50)
143  On Error Goto 2000


160  Rem THE NEXT STATEMENTS CONTROL INPUTS OF PARAMETERS
170  Print" MODEL IS REPRESENTED:"
180  Print"  SURFACE       ADSORBED       COUNTER        DIFFUSE"
190  Print"   PSI(S)        PSI(I)        PSI(C)         PSI(D)"
200  Print"   Psi1          Psi2          Psi3           Psi4"
210   For I=1 To 5
211   Print"    |             |             |              |"
212   Next I
220  Print"      | CAPTNCE Gsa | CAPTNCE Gac| CAPTNCE Gcd |"
230  Print"  SIGMA(S)      SIGMA(I)      GAMMA(NA)      SIGMA(D)"
240  Print"                              GAMMA(CL)"
250  Print"  Sigma1        Sigma2       Gamma3 & 4       Sigma4
255  Print:Print
256  Print  "ADSORPTION IN THE A PLANE GENERATES DIFFUSION INTO THE SURFACE"
```

133

```
260    Print : Print : Print : Print"VALUES OF PARAMETERS"
270    Print"MAX ADSORPTION IN S PLANE IN MICROMOLES/SQ.M ( 10 )   ",Tab(55)
271    Input Maxadss
272    Lprint"MAX ADSORPTION IN S PLANE IN MICROMOLES/SQ.M ",Tab(55);Maxadss
275    Print "CAPACITANCES TO BE ENTERED AS FARADS/SQ.M."
280    Print"CAPACITANCE Gsa( 3 )",Tab(55); : Input Gsa
285    Lprint"CAPACITANCE Gsa( 3 )",Tab(55); Gsa
290    Print"CAPACITANCE Gac( 5 )",Tab(55); : Input Gac
295    Lprint"CAPACITANCE Gac( 5 )",Tab(55);Gac
300    Print"CAPACITANCE Gcd(V. LARGE eg 1e12 or 0.2 )",Tab(55); : Input Gcd
305    Lprint"CAPACITANCE Gcd",Tab(55);Gcd
310    Print"CHECK - CAPTNCE FROM SURFACE TO COUNTER IONS IS";1/(1/Gsa+1/Gac)
312    Print : Print
315    Print"IF YOU ARE WORKING AT CONST IONIC STRENGTH YOU MAY WANT TO USE"
316    Print "CONCENTRATION NOT ACTIVITY.  IF SO, YOU WILL HAVE TO ENTER"
317    Print "DISSOCIATION CONSTS FOR THAT IONIC STRENGTH"
320    Input "CONCENTRATION OR ACTIVITY (C/A)",Act$


330    Print"K(H) ( 1e7 )",Tab(55); : Input Kh
331    Lprint"K(H) ( 1e7 )",Tab(55); Kh
333    Print"K(OH) ( 1e4 )",Tab(55); : Input Koh
334    Lprint"K(OH) ( 1e4 )",Tab(55); Koh
335    Print"K(CAT) ( 1 )",Tab(55); : Input Kcat
336    Lprint"K(CAT) ( 1 )",Tab(55); Kcat
337    Print"K(AN) ( 1 )",Tab(55); : Input Kan
338    Lprint"K(AN) ( 1 )",Tab(55); Kan


350    Print"MAX. ADS IN PLANE A IN MICROMOLES/SQ M. (FOR P 2.5 )",Tab(55);
351    Input Maxada
360    Kdiss(1)=7.58E-03 : Kdiss(2)=6.166E-08 : Kdiss(3)=2.137E-13
365    If Act$="C" Or Act$="c" Then Goto 385
370    Input "IS ADSORBATE PHOSPHATE (Y/N)? ",Phos$
380     If Phos$="Y" Or Phos$="y" Then 399
385    Print"DISS CONSTS FOR ADSORBING SPECIES. ENTER ZEROES FOR 2ND &/OR "
387    Print"3RD CONST IF THEY DO NOT EXIST"
390    Lprint"DISS CONSTS FOR ADSORBING SPECIES. "
391    Print"FIRST",Tab(55); : Input Kdiss(1)
392    Lprint"FIRST",Tab(55); Kdiss(1)
395    Print"SECOND",Tab(55); : Input Kdiss(2)
396    Lprint"SECOND",Tab(55); Kdiss(2)
397    Print"THIRD",Tab(55); : Input Kdiss(3)
398    Lprint"THIRD",Tab(55); Kdiss(3)
399    Print"BINDING CONSTANTS FOR ADSORBING SPECIES. ENTER ZEROES FOR NON "
400    Print"ADSORBED SPECIES. SUGGESTED VALUES ARE FOR P"
```

```
401    Lprint"BINDING CONSTANTS FOR ADSORBING SPECIES."
402    Print"MONOVALENT (0 )";Tab(55); : Input Kaff(1)
403    Lprint"MONOVALENT (0 )";Tab(55); Kaff(1)
405    Print"DIVALENT ( 10) )";Tab(55); : Input Kaff(2)
406    Lprint"DIVALENT ( 10) )";Tab(55); Kaff(2)
407    Print"TRIVALENT ( 0 )";Tab(55); : Input Kaff(3)
408    Lprint"TRIVALENT ( 0 )";Tab(55); Kaff(3)


409    Print "VALUE OF ALPHA ARROW FORWARD ( 2 ) ";Tab(55); :Input Alpharrow
410    Lprint "VALUE OF ALPHA ARROW FORWARD  ( 2 ) ";Tab(55); Alpharrow
411    Print "VALUE OF ALPHA ARROW BACK  ( 0 ) ";Tab(55); :Input Alpharrowb
412    Lprint "VALUE OF ALPHA ARROW BACK  ( 0 ) ";Tab(55); Alpharrowb
413    Print "RATE CONSTANT FOR BACK REACTN (0.15 PER HR)";Tab(55):Input Kback
414    Lprint "RATE CONSTANT FOR BACK REACTION ";Tab(55);Kback
415    Print "DIFFUSION CONST AT 25C (C 0.00003/HR) ";Tab(55);: Input Kdiff25
416    Lprint "DIFFUSION CONSTANT AT 25 Deg ( PER HR) ";Tab(55); Kdiff25
417    Print "ACTIVTN ENERGY FOR DIFFUSION (83000 )";Tab(55);:Input Actenergy
418    Lprint "ACTIVATION ENERGY FOR DIFFUSION ( 83000 )";Tab(55); Actenergy
419    Print"VALENCY OF ELECTROLYTE CATION",Tab(55); : Input Vcat
420    Print"VALENCY OF ELECTROLYTE ANION",Tab(55); : Input Van
421    Lprint"VALENCY OF ELECTROLYTE CATION",Tab(55); Vcat
422    Lprint"VALENCY OF ELECTROLYTE ANION",Tab(55); Van


425    Print : Print" VALUES OF VARIABLES"
428    Print "MASS OF VC OXIDE PER LITRE (G/L)   ",Tab(55);:Input Wt
429    Lprint "MASS OF VC OXIDE PER LITRE (G/L)   ",Tab(55); Wt
431    Print "SURFACE AREA OF THE GOETHITE (SQ M/ G)",Tab(55);:Input Area
432    Lprint "SURFACE AREA OF THE GOETHITE (SQ M/G) ",Tab(55);Area
433    Print "TEMPERATURE IN DEG C  ",Tab(55);:Input Ctemp:Abstemp=Ctemp+273
434    Lprint "TEMPERATURE IN DEG C  ",Tab(55); Ctemp
437    Kbind=Exp(298*Log(Kaff(2)*1e6)/Abstemp)*1e-6
438    Kdiff=Kdiff25*Exp(Actenergy/(298*8.3144) − Actenergy/(Abstemp*8.3144))
439    Print "INITIAL CONC OF ADSORBATE (Micromolar) ",Tab(55);
440    Input Concads : Originalconc=Concads
444    Lprint "INITIAL CONC OF ADSORBATE (Micromolar) ",Tab(55); Concads
445    Dilfactor=1
446    Print"pH",Tab(55); : Input Ph
447    Lprint"pH",Tab(55); Ph
448    Print"ELECTROLYTE CONC.(MOLAR SCALE)",Tab(55); : Input Concelec
449    Lprint"ELECTROLYTE CONC.(MOLAR SCALE)",Tab(55); Concelec


450    Rem SIMPLE CALCULATION OF ACTIVITY COEFF FOLLOWS.
451    Rem ALTER IF YOU WANT MORE COMPLEX VERSION
```

```
455    Rem THIS SIMPLE VERSION IS NOT ACCURATE AT HIGH CONCENTRATIONS
460    Mu=0.5*(Van*Concelec*Vcat^2+Vcat*Concelec*Van^2)
470    Actcoef=10^(-0.5*Sqr(Mu)/(1.0+Sqr(Mu)))
480    Print "Mu , Act ",Mu,Actcoef
485    If Act$="C" Or Act$="c" Then Actcoef=1


490    Rem ADJUST DISSOCIATION CONSTANTS.
500    Kdiss1=Kdiss(1)/Actcoef^2 : Kdiss2=Kdiss(2)/Actcoef^4
505    Kdiss3=Kdiss(3)/Actcoef^6
512    Print : Print


513    Print "TIME FOR FIRST STEP IN HR ",Tab(55);:Input Time(1)
514    Occsites(0)=0: Sumocc=0: Tindex=1: Tnow=Time(1): Tlast=0
515    Sumprod(1)=Time(1)*Kdiff


520    Rem THE NEXT STATEMENTS MAKE A FIRST ESTIMATE OF PSI(S).
530    Conch=10^(-Ph) : Concoh=1E-14/Conch
550    Term=Sqr(Koh*1E-14/Kh) : Zpc=-Log(Term)/Log(10)
560    Psi1=58*(Zpc-Ph)
570    Print Using"NERNST EST  ####.#";Psi1


580    Rem THE NEXT STATEMENTS ARE THE SET OF CHARGING EQUATIONS
590    Rem THEY END AT 980 WITH A SECOND ESTIMATE OF SIGMA)(S) - Estsigma1
600    Print
610    Delta1=10:Ii=0
620    I=1


625    Rem CALC. CHARGE IN FIRST LAYER INCLUDING THAT DUE TO PENTRTD ADSORBATE
630    Hterm=Kh*Conch*Exp(-0.039*Psi1*298/Abstemp)
640    Ohterm=Koh*Concoh*Exp(0.039*Psi1*298/Abstemp)
641    Gammah=Maxadss*Hterm/(1+Hterm+Ohterm)
642    Gammaoh=Maxadss*Ohterm/(1+Hterm+Ohterm)
643    If Tindex=1 Then Fixed=0 : Goto 650
644    Sumterm=0
645     For K=2 to Tindex
646     Factor=1-Sumocc/Maxada
647     Sumterm=Sumterm+Occsites(K-1)*Sqr((Sumprod(K)/Factor))
648     Next K
649    Fixed=Sumterm*1.12838:    Rem  2/ROOT PI
650    Deltafixed=Fixed-Oldfixed
653    Concads=(Originalconc-(Fixed+sumocc-Deltafixed)*Wt*Area)/Dilfactor
655    Sigma1=Gammah-Gammaoh-2.0*Fixed
```

```
658    Rem  CALC PSI IN SECOND LAYER THEN CHARGE
660    Psi2=Psi1-Sigma1*96.487/Gsa
661    Rem USE FARADAY CONSTANT TO GIVE POTNL IN MILLIVOLTS
670    If Concads=0 Then Sigma2=0 : Goto 770
680    T1=Conch^3+Conch^2*Kdiss1+Conch*Kdiss1*Kdiss2+Kdiss1*Kdiss2*Kdiss3
740    Alpha=Conch*Kdiss1*Kdiss2/T1

745    Rem NOTE -SET FOR DIVALENT IONS. 740 & 752 MUST BE CHNGD FOR OTHER IONS

748    M0=(Maxada-Sumocc+Deltafixed)*Wt*Area/Dilfactor
749    S0=(Sumocc-Deltafixed)*Wt*Area/Dilfactor
750    C0=Concads
751    Eterm=Exp(Alpharrow*Psi2*0.039*298/Abstemp)
752    K1star=Kbind*Kback*(Alpha*Actcoef^4)*Eterm
753    K2star=Kback*Exp(-Alpharrowb*Psi2*0.039*298/Abstemp)
754    A=C0-M0-K2star/K1star
756    B=(C0+S0)*K2star/K1star
758    Rt=Sqr(A^2 + 4*B)
760    Ce=(A+Rt)/2: Q=(Rt-A)/2
761    Eterm=(Ce+Q)*K1star*Time(Tindex)
762    If Eterm>80 Then Ct=Ce:Goto 766
764    Capb=Exp(Eterm)*(C0+Q)/(C0-Ce)
765    Ct=(Q+Capb*Ce)/(Capb-1)
766    Occsites(Tindex)=(C0-Ct)*Dilfactor/(Wt*Area)-Deltafixed
768    Estsumocc=0
770     For K=1 to Tindex: Estsumocc=Estsumocc+Occsites(K): Next K
775    Sigma2=-2.0*Estsumocc


780    Rem NOW CALC PSI & THEN CHARGE IN THIRD LAYER
790    Psi3=Psi2-(Sigma1+Sigma2)*96.487/Gac
800    Rem  CALC MOLARITY OF INDIVIDUAL IONS FOR NON SYMMETRICAL ELECTROLYTE.
810    Conccat=Concelec : If Van>Vcat Then Conccat=Concelec*Van
820    Conccat=Conccat*Actcoef^(Vcat^2)
840    Concan=Concelec : If Vcat>Van Then Concan=Concelec*Vcat
850    Concan=Concan*Actcoef^(Van^2)
860    Anionterm=Kan*Concan*Exp(0.039*Van*Psi3*298/Abstemp)
865    Cationterm=Kcat*Conccat*Exp(-0.039*Vcat*Psi3*298/Abstemp)
880    Gamma4=Maxadss*Van*Anionterm/(1+Anionterm+Cationterm)
890    Gamma3=Maxadss*Vcat*Cationterm/(1+Anionterm+Cationterm)


895    Rem CALC PSI & THEN CHARGE FOT THE DIFFUSE CHARGE
900    Psi4=Psi3-(Sigma1+Sigma2+Gamma3-Gamma4)*96.487/Gcd
910    Term=0.0195*298/Abstemp
930    If Vcat=Van Then 931 else 940
931    T5=Vcat*Term*Psi4
932    If Abs(T5)>87 Then Sigma4=-Sgn(Psi4)*1e35: Goto 980
```

```
934    T6=0.5*(Exp(T5)-Exp(-T5))
936    Sigma4=-1.22*Sqr(Concelec)*T6
938    Goto 980
939    Term=0.0195*298/Abstemp
940    X1=2.0*Term*Psi4*Vcat : X2=2.0*Term*Psi4*Van
945    If Abs(X1)>87 Then Sigma4=-Sgn(Psi4)*1e35: Goto 980
950    W3=(Exp(-X1)-1)/Vcat : W4=(Exp(X2)-1)/Van
960    W5=Sqr(W3+W4) : If Psi4>0 Then W5=-W5
970    Sigma4=.61*Sqr(Concelec)*W5
980    Estsigma1=-Sigma2-Gamma3+Gamma4-Sigma4
985    Rem  END OF CALCULATION OF CHARGING EQUATIONS


990    Rem  NOW COMPARE THE 2 ESTIMATES OF SIGMA(S) & ADJUST PSI(S)
1000   Difference=Sigma1-Estsigma1
1005   If Abs(Difference)<1E-04 Then Goto 1140
1006   Delta2=Sgn(Difference)*Delta1


1007   Rem IF DIFFERENCE LARGE USE INCREMENTAL SEARCH
1008   If Abs(Difference)>100 Goto 1110


1009   Rem WHEN DIFFERENCE SMALL ENOUGH USE LINEAR EXTRAPOLATION
1010   If I=1 Then Prepsi=Psi1 : Goto 1090
1020   Slope=(Difference-prediff)/(Prepsi-Psi1)
1030   Intercept=Difference+Slope*Psi1
1035   Prepsi=Psi1
1037   If Slope=0 then Psi1=Prepsi*0.999: Goto 1090
1040   Psi1=Intercept/Slope
1050   If Psi1=Prepsi1 Then Psi1=Psi1*0.999
1090   Prediff=Difference
1091   If I=11 Then 1092 Else 1094
1092   Print "SLOW CONVERGENCE. PRINT OUT SUCCESSIVE ESTS OF  POTNL"
1093   Print" & OF DIFFERENCE"
1094   If I>10 Then Print I,Prepsi,Difference
1095   If I>20 And Abs(Difference)<1e-3 Goto 1140
1096   If I=30 Then Print "NOGO! BETTER RECONSIDER THE PARAMETERS":Goto 1280
1100   I=I+1
1105   If I=2 Then Psi1=Psi1+Delta2
1106   Goto 630
1108   Rem END OF THE EXTRAPOLATION & THE BEGINNING OF THE INCREMENTAL SEARCH


1110   Ii=Ii+1: If I=1 Then I=2
1115   If Ii=1 Goto 1128
1125   Prod=Sgn(Difference)*Sgn(Prediff)
```

```
1126    If Prod<0 Then Delta1=0.49*Delta1
1128    Prepsi=Psi1:Psi1=Psi1+Delta2
1129    Prediff=Difference
1130    Goto 630
1132    Rem END OF INCREMENTAL SEARCH


1140    Rem THE NEXT STATEMENTS CONTROL OUTPUT.
1150    Print Tab(31);"SURFACE        ADSORBED       COUNTER        DIFFUSE"
1160    Print"POTENTIAL (M.VOLTS)       ";
1165    Print Using"      ####.###";Psi1,Psi2,Psi3,Psi4
1170    Print"CHARGE (MICRO M/SQ.M.)   ";
1175    Print Using"         ##.###";Sigma1,Sigma2,Gamma3-Gamma4,Sigma4
1180    Print Tab(23):Print Using"GAMMA H   ##.###";Gammah;
1181    Print Using"          GAMMA CAT. ##.###";Gamma3
1182    Print Tab(23):Print Using"GAMMA OH  ##.###";Gammaoh;
1183    Print Using"          GAMMA AN.  ##.###";Gamma4
1185     If Sp$="Y" Or Sp$="y" Goto 1240
1190    Lprint Tab(31);"SURFACE        ADSORBED       COUNTER        DIFFUSE"
1200    Lprint"POTENTIAL (M.VOLTS)       ";
1202    Lprint Using"      ####.###";Psi1,Psi2,Psi3,Psi4
1205    Lprint"CHARGE (MICRO M/SQ.M.)   ";
1206    Lprint Using"         ##.###";Sigma1,Sigma2,Gamma3-Gamma4,Sigma4
1210    Lprint Tab(23):Lprint Using"GAMMA H   ##.###";Gammah;
1215    Lprint Using"          GAMMA CAT. ##.###";Gamma3
1220    Lprint Tab(23):Lprint Using"GAMMA OH  ##.###";Gammaoh;
1225    Lprint Using"          GAMMA AN.  ##.###";Gamma4


1240    Print
1242    Sumocc=Estsumocc
1245    Print Using"TOTAL SORPTION (MICRO M/SQ.M.)        ##.######";Sumocc+Fixed
1246    Print Using"TRUE ADSORTION                        ##.######";Sumocc
1247    Print Using"PENETRATED                            ##.######";Fixed
1248    If Sp$="Y" Or Sp$="y" Goto 1260
1250    Lprint Using"TOTAL SORPTION (MICRO M/SQ.M.)        ##.######";Sumocc+Fixed
1252    Lprint Using"TRUE ADSORTION                        ##.######";Sumocc
1255    Lprint Using"PENETRATED                            ##.######";Fixed
1260    Print  :Print Using"NO OF OUTER CYCLES   ##";Ii
1262    Print Using"NO OF EXTRAPOLATIONS ##";I: Print : Print
1263    Oldfixed=Fixed
1265    Concads=(Originalconc-(Sumocc+Fixed)*Wt*Area)/Dilfactor
1266    Print"CURRENT CONC IS    ",Concads;" AFTER A PERIOD OF ",Tnow
1267    Lprint"CURRENT CONC IS    ",Concads;" AFTER A PERIOD OF ",Tnow
1268    Tlast=Tnow


1270     Input " SHALL WE SKIP PRINTER ON NEXT TIME CYCLE? (Y/N) ",Sp$
```

```
1280    Print" TO ALTER TEMP, ENTER 1"
1283    Print" TO ALTER DILUTION &/or pH, ENTER 2"
1285    Print" TO ALTER DIFF COEF, ENTER 3"
1287    Print" TO ALTER pH OR CONC, ENTER 4
1290    Print" TO CONTINUE, ENTER 5"
1297    Print" TO QUIT ENTER 6"
1298     Print" TO ALTER VALUE FOR KBACK ENTER 7"
1300    Input Index
1310    On Index Goto 1350,1400,1450,439,1500,1900,1800


1350    Input "NEW TEMP.  ",Ctemp: Abstemp=Ctemp+273
1355    Lprint"NEW TEMP.  ",Ctemp
1360    Kbind=Exp(298*Log(Kaff(2)*1e6)/Abstemp)*1e-6
1370    Kdiff=Kdiff25*Exp(Actenergy/(298*8.3144) - Actenergy/(Abstemp*8.3144))
1375    Goto 1280


1400    Input "DILUTION FACTOR ", Dilfactor
1405    Lprint : Lprint : Lprint "DILUTION FACTOR  ",Dilfactor
1410    Concads=Concads/Dilfactor
1425    Input "NEW pH  ",pH
1427    Lprint "NEW pH  ",pH
1428    Conch=10^(-Ph) : Concoh=1E-14/Conch
1429    Psil=58*(Zpc-Ph)
1430    Goto 1280


1450    Input "NEW VALUE FOR Kdiff25  ",Kdiff25
1455    Lprint "NEW VALUE FOR Kdiff25    ",Kdiff25
1460    Kdiff=Kdiff25*Exp(Actenergy/(298*8.3144) - Actenergy/(Abstemp*8.3144))
1470    Goto 439


1500    Input  "NEW TOTAL TIME    ",Tnow
1510    Lprint "NEW TOTAL TIME    ",Tnow
1520    Tindex=Tindex+1:  Time(Tindex)=Tnow-Tlast
1530    Coefdiff(Tindex)=Kdiff
1540     For J=1 To Tindex  :Sumprod(J)=0  : Next J
1550     For J=1 To Tindex
1560      For Jj=J to Tindex
1570       Sumprod(J)=Sumprod(J)+Time(Jj)*Coefdiff(Jj)
1580      Next Jj
1590     Next J
1600    Goto 600

1800     Input"NEW VALUE FOR K2  ",Kback
1810     Lprint : Lprint : Lprint "NEW VALUE FOR K2 ",Kback
```

```
1820    Goto 1280
1900    End


2000 Print "ERROR NO",Err,"  AT LINE NO ",Erl : Resume Next
```

Chapter B5

Fitting models to data

In Chapters B1, B2 and B4 some examples involving iteration were considered. This Chapter considers a third kind of iteration. Suppose we have a model which we think is appropriate and a method of solving the equations that constitute the model. Suppose further that we have a set of data and we want to test whether the model can, in fact, describe that data. The data might consist of a single response variable (or dependent variable) — such as the amount of adsorption. Or there might be more than one response variable — such as, the amount of adsorption, the charge on the surface, and the zeta potential. There might also be several explanatory (or independent) variables — such as the pH, the concentration of adsorbate, and the ionic strength of the medium. The problem is then to find appropriate values of the parameters. Examples of parameters are: the maximum adsorption possible for a given ion; the capacitance of the regions between the mean planes of adsorption; and the binding constants for the individual ions. The solution to this problem depends on the modellers' definition of the word "appropriate". He might argue that the parameters should be chosen from independent knowledge of the system — for example, the maximum adsorption from the geometry of the surface and of the adsorbing ion; the capacitance from the distances between the layers and from the dielectric properties; and the binding constants from knowledge of the bonds formed. An alternative argument is that the knowledge for this ideal solution is not adequate. Hence the appropriate values are those which best tune the model to the data. In this case the level of knowledge required is obviously less and the hypothesis being tested is whether the model is indeed capable of describing the data. (Of course a mix of these strategies could also be considered. The values of some of the parameters might be fixed from independent knowledge.) Before considering the problem further, it is desirable to consider the meaning of "best" when tuning the model to the data. A common criterion against which to choose the best is that the sums of squares for the deviations be as small as possible — and this is the criterion used here. However other criteria can be used (Bard 1974). Further it is important that parts of the data set do not unduly influence the result. Thus if the precision decreases as the value of the response variable increases, high values will have an inappropriately large effect. In this case the data should be transformed. If, for example, a log. transformation were appropriate, the sums of squares minimised would be that for the difference between the log. of the observed value and the log. of the predicted value. Finally if there is more than one response variable, it may be desirable to weight the sums of squares so that one of the response variables does not swamp the others.

A program to handle this "model-fitting" job conveniently has three components. These are:

1) A main program which reads in the data, plus a first estimate of the values of the parameters. It then hands control to a sub-program which varies the values of the parameters. Subsequently, when the best values of the parameters have been found, control is handed back to the main program and the results are printed.

2) The subprogram "SIMPLEX" uses the simplex method of Nelder and Mead (1965) to control the variations of the parameters. A good discussion of this method will be found in Olsson and Nelson (1975). However, for the present purposes, it suffices to regard it merely as a program that twists the knobs until it finds the best setting.

3) The "SIMPLEX" subprogram calls the further subprogram "EQUATION" in which

142

the model is specified and which also specifies the property to be minimized (called "F" in the code). The minimum value of F defines the best values of the parameters.

This modular arrangement has several advantages when a diverse range of models is being considered. The data to be read in will vary somewhat between jobs — but this can be accommodated by appropriate changes in the input statements. Hence only small changes will be needed in the main program. The SIMPLEX subprogram should be equally applicable in all cases and hence should not need any changes. All that needs to be done is to graft on the appropriate "EQUATION" subroutine. Thus in subsequent listings in this book, the "SIMPLEX" subroutine will not be repeated. The user will have to merge it with the MAIN and EQUATION routines. The listing given here is appropriate to fit the four-plane model to data for phosphate adsorption on goethite. As written, the program is appropriate if final solution concentration of phosphate is to be predicted from the initial concentration of the pH. However "Rem" statements indicate how to change the program in order to predict adsorption from the final concentration and to show how to predict charge. The program may also be used to fit the measurements of charge in the absence of adsorption. It is also possible to permit all parameters to vary and to fit simultaneously to data for charge and for adsorption.

References

Bard, Y. 1974. Nonlinear parameters estimation. Academic Press, New York.

Nelder, J.A. and Mead, R. 1965. A simplex method for function minimisation. Computer Journal 7, 308-313.

Olsson, D.M. and Nelson, L.S. 1975. The Nelder-Mead Simplex procedure for function minimisation. Technometrics 17, 85-51.

```
1    Rem                    FITPINIT.BAS


10    Print"THIS IS THE PGME TO FIT THE FOUR PLANE MODEL TO P ADSORPTION."
15    Print"IT USES THE INITIAL CONC AS INPUT & PREDICTS FINAL CONC."
16    Print"CHANGES TO MAKE IT WORK USING FINAL CON & PREDICTING ADSORPTION"
17    Print"ARE INDICATED"
20    Print"THE BACKGROUND DATA ON THE GOETHITE IS READ IN AS 'Data'"
26    Print"THESE WOULD HAVE BEEN ALLOCATED TO FIT TITRATION DATA USING A"
27    Print"SIMILAR PROGRAM. IT IS ALSO POSSIBLE TO FIT SIMULTANEOUSLY TO"
28    Print"ADSORPTION DATA & TO TITRATION DATA THUS FITTING ALL PARAMETERS."
29    Print"THE CONTROL NUMBERS SUCH AS NO OF OBS. ARE ALSO READ AS Data"
30    Print"THE DATA TO BE FITTED ARE READ FROM A FILE."
40    Print"THE ALLOCATION OF THE VARIABLES & OF THE PARAMETERS IS SPECIFIED"
50    Print"IN STATEMENTS IN THE SUBROUTINE BEGINNING AT 4000 ."




70    Time$="0:0"
80    Data 84:Rem              AREA IN METER SQ PER G
90    Data 10: Rem             MAX ADS IN S PLANE IN MICROMOLES/SQ M
100   Data 2.5:Rem             MAX ADSORPTION IN A PLANE IN MICROMOLES/SQ M
110   Data 2.0,1E10:Rem        CAPACITANCE G-s-beta & G-beta-d IN F/SQ M
120   Data 4.3e7,5e3:Rem       K(H) & K(OH)
130   Data 0.75,.05:Rem        K(CAT) & K(AN)
140   Data .1:Rem              CONC OF ELECTROLYTE
145   Data 2.1 : Rem           WT OF OXIDE PER LITRE
150   Data 1:Rem               IF THIS 1, DISSN CONSTS FOR P ARE ASSUMED,
151   Rem                      IF NOT, THE CONSTS MUST BE READ IN
160   Rem Data 0,0,0:Rem       IF  PREVIOUS NO=0 READ IN DISSN CONSTS HERE
170   Data 1,1 :Rem            VALENCY OF ELECTROLYTE CATION & ANION
180   Data XXXXX:Rem           NAME OF FILE FOR DATA
190   Data 50: Rem             NO OF OBSERVATIONS
200   Data 3:Rem               NO OF PARAMETERS TO BE VARIED
210   Data 2.5,14,3: Rem       FIRST ESIMATES OF THE PARAMETERS


220   On Error Goto 4940

230   Dim Y(80),Ye(80),X(3,80),Estchg(80),Psis(80),Param(10),Title$(100)
240   Dim Act(4),Kaff(4),Sumchg(4),Sumads(4),Concion(4),Kdiss(4)
```

```
250     Lprint :Lprint"VALUES OF PARAMETERS":Lprint:Lprint
260     Restore
265     Print "INPUT FROM DATA STATEMENTS ARE BEING LISTED TO THE PRINTER"
270     Read Area:Lprint "SURFACE AREA OF GOETHITE (SQ M/G)          ",Area
280     Read Ns:Lprint "VALUE FOR N(S)                               ",Ns
290     Read Maxads :Lprint "MAXADS IN THE A PLANE                    ",Maxads
300     Read Gsc :Lprint "CAP BETWEEN S & BETA  PLANES IN FARADS/SQ.M ", Gsc
310     Read Gcd :Lprint "CAP BETWEEN BETA & D PLANES IN FARADS/SQ.M", Gcd
320     Read Kh :Lprint "Kh                    ",Kh
321     Read Koh : Lprint"Koh                   ", Koh
330     Read Kcat :Lprint "AFFINITY TERM FOR CATIONS                  ",Kcat
340     Read Kan:Lprint "AFFINITY TERM FOR ANIONS                     ",Kan
350     Read Concelec
351     Lprint "CONC OF 1:1 ELECTROLYTE                       ",Concelec
355     Read Wt:Lprint "WT OF OXIDE PER LITRE                        ",Wt


360     Rem THE FOLLOWING SIMPLE ALLOCATION OF ACT COEFS IS ADEQUATE FOR
361     Rem MANY PURPOSES. IF NOT CHANGE IT!
370     If Concelec=0.01 Then Actcoef=0.903
380     If Concelec=0.1 Then Actcoef=0.778
390     If Concelec=1 Then Actcoef=0.658

400     Rem IN THIS CASE ELECTROLYTE CONC WAS CONSTANT -SO WAS READ IN AS DATA
410     Rem IF IT VARIED IT WOULD BE READ IN FROM THE FILE FOR EACH TREATMENT
415     Rem AND ACTIVITY WOULD BE CALCULATED & ADJUSTED FOR IN THE SUBROUTINE
440     Kdiss(1)=7.58E-03 : Kdiss(2)=6.166001E-08 : Kdiss(3)=2.137E-13
455     Kdiss1=Kdiss(1)/Actcoef^2 : Kdiss2=Kdiss(2)/Actcoef^4
456     Kdiss3=Kdiss(3)/Actcoef^6
450     Read L:If L=1 Then Lprint "ADSORBATE IS PHOSPHATE"
451     Rem IF NOT PHOSHATE MAKE L=0
460     If L=1 Then Goto 530


470     Lprint"DISS CONSTS FOR ADSORBING SPECIES. ZEROES TO BE ENTERED FOR "
480     Lprint"2nd &/OR 3rd CONST IF THEY DO NOT EXIST"
490     Read Kdiss1:Lprint"FIRST",TAB(24);  Kdiss1
500     Read Kdiss2:Lprint "SECOND",TAB(24);  Kdiss2
510     Read Kdiss3:Lprint"THIRD",TAB(24);  Kdiss3
530     Kaff(1)=0 : Kaff(2)=0 : Kaff(3)=0


540     Read Vcat:Lprint "VALENCY OF ELECTROLYTE CATION  ",Vcat
550     Read Van:Lprint "VALENCY OF ELECTROLYTE ANION   ",Van
560     Read Nname$:Lprint "NAME OF FILE FOR Data     ";Nname$
```

```
570    Rem  NOW OPEN FILE TO INPUT THE OBSERVATIONS
580    Open "B:"+Nname$ For Input As #1
590    Input #1,Title$
600    Lprint : Lprint : Lprint Title$
610    Read Nobs:Lprint "No OF POINTS        ",Nobs
620      For I=1 To Nobs
630      Input #1,Y(I),X(1,I),X(2,I)
635      Rem MODIFY IF MORE VARIABLES ARE INVOLVED. NOTE THAT X(1, ) IS pH.
640      Lprint I,Y(I),X(1,I),X(2,I)
660      Next I
670    Lprint


680    Lprint"PARAMETERS ARE AS SPECIFIED IN SUBROUTINE EQUATION    "
690  . Read Nparam:Lprint "NUMBER OF PARAMETERS        ",Nparam
700    Lprint"INITIAL ESTIMATES OF PARAMETERS"
710      For I=1 To Nparam
720      Lprint"PARAMETER  ",I; : Read  Param(I):Lprint  Param(I)
730      Next I


740    Rem    NOW MAKE INITIAL ESTIMATES OF PSIS FOR STARTING POINT
750    Zpc=-(0.5/2.303)*(Log(Koh*1E-14)-Log(Kh))
760      For I=1 To Nobs : Psis(I)=58*(Zpc-X(1,I)) : Next I

770    Nmax=300 : Tolrnce=0.01 : Rem ADJUST THESE VALUE TO SUIT YOUR JOB
775    Rem DECREASE Tolrnce IF IT JUMPS OUT BEFORE THE MIN IS STABLE


780     Print"VALUES FOR PARAMETERS, VALUE OF FUNCTION BEING MINIMISED "
781     Print"& LAPSED TIME WILL BE TABULATED ON SCREEN"


790    Rem TRANSFER TO SIMPLEX TO OPTIMISE PARAMETERS
800    Gosub 2030


810    Rem  RETURN TO EQUATION ROUTINE TO USE VALUES FROM BEST VERSION
820    Gosub 4000
```

```
830   Rem    EVALUATE REGRESSION PARAMETERS & PRINT OUT RESULTS
840   Ysum=0 : Ysumsq=0 : Devsq=0
850     For I=1 To Nobs
860     Ysum=Ysum+Log(Y(I)):Rem  ASSUMES WE ARE FITTING TO LOG TRANS VALS
870     Ysumsq=Ysumsq+Log(Y(I))^2
880     Devsq=Devsq+(Log(Y(I))-Log(Ye(I)))^2
890     Next I
900   Sumssqy=Ysumsq-Ysum^2/(Nobs)
910   R1=1.0-Devsq/Sumssqy



920   Lprint:Lprint:Lprint "JOB COMPLETED":Lprint:Lprint
930   Lprint : Lprint"R SQ            ",R1
940   Lprint : Lprint
950   Lprint"TOTAL SUMS OF SQUARES",SUMSSQY
960   Lprint "RESID SUMS OF SQUARES",DEVSQ
965   Lprint "VALUE OF FUNCTION MINIMISED ",F
970   Lprint : Lprint


980   Lprint"VALUES OF PARAMETERS"
990   Lprint
1000    For I=1 To Nparam
1010    Lprint Using"PARAMETER ##";I;: Lprint"     ";: Lprint Param(I)
1020    Next I
1030  Lprint : Lprint
1040  Lprint"PLOTTING POINTS"
1050  Lprint : Lprint TITLE$ : Lprint

1055  Rem CHANGE THE NEXT SECTION IF THE JOB IS DIFFERENT
1060  Lprint"            NO           Y          PREDICTED   Y       DEV";
1061  Lprint Tab(79);:Lprint "pH     X          PRED ADS"
1070    For I=1 To Nobs
1080    Lprint Using"    ##.#####";I,Y(I),Ye(I),(Log(Y(I))-Log(Ye(I)));
1090    Lprint Using"    ####.###";X(1,I),X(2,I),(X(2,I)-Ye(I))/(Area*Wt)
1100    Lprint
1110    Next I
1120  Lprint
1130  Lprint Using"NUMBER OF CYCLES  ####";Nevals
1140  Lprint
1150  Lprint Using"TIME LAPSE \         \";Time$
1160  Lprint"FINISH"
1170  End
```

```
2030    Rem              SIMPLEX


2040    Rem THIS SUBROUTINE SEARCHES FOR A MINIMUM VALUE OF "F"
2050    Rem THE SUBROUTINE "EQUATION" SPECIFIES F.


2060    Dim Ppoint(10,10),Fpoint(10),Pointo(10)
2070    Dim Delparam(10)
2080    N=Nparam


2090    Rem SET UP ARRAY OF FIRST STEPS
2100      For I=1 To N
2110      Delparam(I)=0.1*Param(I)
2120      Next I
2140    Nevals=0
2150    Ncnfirm=Nmax/20
2160    N1=N+1
2170    Recipn=1.0/N


2175    Rem SET UP INITIAL SIMPLEX OF N+1 POINTS & EVALUATE F AT EACH POINT
2180      For I=1 To N1
2190        For J=1 To N
2200        Ppoint(I,J)=Param(J)
2210        Next J
2220      Next I
2230      For I=2 To N1
2240      Ppoint(I,I-1)=Param(I-1)+Delparam(I-1)
2250      Next I
2260      For I=1 To N1
2270        For J=1 To N
2280        Param(J)=Ppoint(I,J)
2290        Next J
2300      Gosub 4000:Rem TO FIND VALUE OF F
2310      Fpoint(I)=F
2320      Next I


2325    Rem FIND BEST & WORST POINTS, THEN COME BACK TO FIND CENTROID
2326    Rem OF ALL POINTS EXCEPT THE WORST
2330    Goto 2870
2340      For J=1 To N
```

```
2350      Pointo(J)=0
2360      Next J
2370      For I=1 To N1
2380      If I=Iworst Then Goto 2420
2390        For J=1 To N
2400        Pointo(J)=Pointo(J)+Ppoint(I,J)*Recipn
2410        Next J
2420      Next I


2425   Rem REFLECT WORST THRU CENTROID TO POINT R & EVALUATE F THERE
2430      For J=1 To N
2440      Param(J)=Pointo(J)+(Pointo(J)-Ppoint(Iworst,J))
2450      Next J
2460   If Nevals>Nmax Then Goto 3090
2470   Gosub 4000


2475   Rem IF RELECTION SUCCESSFUL, REPLACE WORST POINT BY R.
2480   If F>Fworst Then Goto 2710
2490      For J=1 To N
2500      Ppoint(Iworst,J)=Param(J)
2510      Next J
2520   Fpoint(Iworst)=F


2525   Rem IF REFLECTION SUCCESSFUL & R IS NOW THE BEST, TRY EXPANSION
2526   Rem IF IT WAS ONLY JUST SUCCESSFUL(ie R IS NOW WORST) TRY CONTRACTION
2527   Rem IF R BETWEEN BEST & WORST GO STRAIGHT TO PREPARE FOR NEXT CYCLE
2530   If F<Fbest Then  Goto 2590
2540      For I=1 To N1
2550      If F<=Fpoint(I) Then Goto 2870
2560      Next I
2570   Fworst=F
2580   Goto 2710


2585   Rem IF REFLECTION WAS VERY SUCCESSFUL, EXPAND SIMPLEX
2586   Rem FURTHER IN DIRECTION TO POINT E
2587   Rem IF THIS IS SUCCESSFUL REPLACE WORST POINT BY E, THEN
2588   Rem PREPARE FOR NEXT CYCLE
2590   Fbest=F
2600      For J=1 To N
2610      Param(J)=Param(J)+2*(Param(J)-Pointo(J))
2620      Next J
2630   If Nevals>Nmax Then Goto 3090
```

149

```
2640    Gosub 4000
2650    If F>Fbest Then Goto 2870
2660      For J=1 To N
2670      Ppoint(Iworst,J)=Param(J)
2680      Next J
2690    Fpoint(Iworst)=F
2700    Goto 2870


2705    Rem IF REFLECTION WAS UNSUCCESSFUL, OR ONLY JUST SUCCESSFUL,
2706    Rem CONTRACT WORST POINT TOWARDS CENTROID TO POINT C.
2706    Rem IF THIS IS SUCCESSFUL, REPLACE WORST POINT BY C.
2710      For J=1 To N
2720      Param(J)=0.5*(Pointo(J)+Ppoint(Iworst,J))
2730      Next J
2740    If Nevals>Nmax Then Goto 3090
2750    Gosub 4000
2760    If F<=Fworst Then Goto 2660


2765      Rem IF CONTRACTION UNSUCCESSFUL, SHRINK EACH OTHER POINT HALF WAY
2766      Rem TOWARDS BEST & EVALUATE F AT THESE NEW POINTS
2770      For I=1 To N1
2780      If I=Ibest Then Goto 2860
2790        For J=1 To N
2800        Ppoint(I,J)=0.5*(Ppoint(Ibest,J)+Ppoint(I,J))
2810        Param(J)=Ppoint(I,J)
2820        Next J
2830      If Nevals>Nmax Then Goto 3090
2840      Gosub 4000
2850      Fpoint(I)=F
2860      Next I


2865    Rem IN ALL CASES RE-DETERMINE BEST & WORST POINTS OF CURRENT SIMPLEX
2866    Rem BEFORE TESTING FOR END OF SEARCH. IF NOT END REPEAT CYCLE.
2870    Ibest=1
2880    Iworst=1
2890    Fbest=Fpoint(1)
2900    Fworst=Fpoint(1)
2910      For I=2 To N1
2920      If Fpoint(I)>=Fbest Then Goto 2950
2930      Ibest=I
2940      Fbest=Fpoint(I)
2950      If Fpoint(I)<=Fworst Then Goto 2980
2960      Iworst=I
2970      Fworst=Fpoint(I)
```

```
2980     Next I


2985     Rem   TEST FOR END OF SEARCH
2990     Fspread=Fworst-Fbest
3000     If Fspread<=Tolrnce Then Goto 3030
3010     Nlosprd=0
3020     Goto 2340
3030     If Nlosprd>0 Then Goto 3050
3040     Fbesto=Fbest
3050     Nlosprd=Nlosprd+1
3060     If Nlosprd<Ncnfirm Then Goto 2340
3070     Fimprve=Fbesto-Fbest
3080     If Fimprve>Tolrnce Then Goto 3010


3085     Rem END OF SEARCH OR MAX NO OF CYCLES. RECORD BEST VALUES & RETURN
3090       For J=1 To N
3100       Param(J)=Ppoint(Ibest,J)
3110       Next J
3120     F=Fbest
3130     Return


4000     Rem                    EQUATION


4010     Rem   PRINT OUT SUCCESSIVE VALUES ON THE SCREEN AS A CHECK
4020     Print Nevals+1,
4030       For Nnn=1 To Nparam : Print Param(Nnn), : Next Nnn


4040     Rem   ALLOCATE MEANINGS TO THE PARAMETERS
4041     Rem  - ADJUST ACCORDING TO THE MODEL YOU WANT TO FIT
4050       Maxads=Param(1) : Kaff(2)=Param(2) : Gsa=Param(3)


4060       Rem  IF CERTAIN VALUES ARE 'ILLEGAL' DISCOURAGE  SIMPLEX FROM THEM
4070     If Maxads<0 Or Kaff(2)<0 Or Gsa<Gsc  Then F=3*Fbest : Return
```

```
4080    F=0
4085    Concsites=Maxads*Wt*Area


4090    Rem CALCULATE CAPACITANCE BETWEEN A & C PLANES
4100    Gac=1.0/(1.0/Gsc-1.0/Gsa)


4110    Rem BEGIN THE CYCLE OF CALCULATIONS FOR EACH SET OF OBS
4120      For J=1 To Nobs
4130      Hplus=10^(-X(1,J))
4140      Ohminus=1E-14/Hplus
4150      C2=X(2,J)
4160      If Nevals<4.Then D1=10 Else D1=2
4170      Icount=1 : Ii=0


4175      Rem   RECYCLE POINT FOR EQUATIONS
4180      Khstar=Kh*Hplus*Exp(-0.039*Psis(J))
4190      Kohstar=Koh*Ohminus*Exp(0.039*Psis(J))
4200      Gammah=Ns*Khstar/(1+Khstar+Kohstar)
4210      Gammaoh=Ns*Kohstar/(1+Khstar+Kohstar)
4220      Sigma1=Gammah-Gammaoh
4230      Psi2=Psis(J)-Sigma1*96.487/Gsa


4232      Rem CALC ACT. OF IONS
4235      Concads=X(2,J)
4240      If Concads=0 Then Sigma2=0 : Gammaa=0 : Goto 4430
4250      Term=Hplus^3+Hplus^2*Kdiss1+Hplus*Kdiss1*Kdiss2+Kdiss1*Kdiss2*Kdiss3
4260      Concion(1)=Hplus^2*Kdiss1/Term
4270      Act(1)=Concion(1)*Actcoef
4280      Concion(2)=Hplus*Kdiss1*Kdiss2/Term
4290      Act(2)=Concion(2)*Actcoef^4
4300      Concion(3)=Kdiss1*Kdiss2*Kdiss3/Term
4310      Act(3)=Concion(3)*Actcoef^9


4315    Rem   THE NEXT SECTION CALCULATES THE CHARGE DUE TO ASORPTION WHEN
4316    Rem   INITIAL CONC IS SPECIFIED
4320    Sigma2=0 : Sumkstar=0
4325    If Concads=0 Then Suma=0 : Gammaa=0 : Goto 4430
4327    Rem   TO SWITCH TO FINAL CONC, ACTIVATE THE FOLLOWING STATEMENT
4328     Rem   Goto 4415
4330     For K=0 To 3
4340      Z=-K
4350      Kstar(K)=Kaff(K)*Act(K)*Exp(-Z*0.039*Psi2)
```

```
4355      Sumkstar=Sumkstar+Kstar(K)
4360       Next K
4365      B=Concads-Concsites-1.0/Sumkstar
4370      C=Concads/Sumkstar
4375      Q=(B+Sqr(B^2+4*C))/2
4380      Suma=(Concads-Q)/(Wt*Area)
4385      For K=1 to 3
4390       Z=-K
4400       Adsion(K)=Suma*Kstar(K)/Sumkstar
4405       Sigma2=Sigma2+Z*Adsion(K)
4410       Next K


4411      Rem  THE NEXT SECTION CAN BE ACTIVATED TO CALCULATE ADSORPTION WHEN
4412      Rem   FINAL CONC RATHER THAN INITIAL CONC IS SPECIFIED.
4413      Rem   TO USE IT, INACTIVATE THE NEXT STATEMENT
4414      Goto 4430
4415       Sumchg=0 : Sumads=0
4417        For K=1 To 3
4419        Ads(K)=Kaff(K)*Concads*Act(K)*Exp(K*0.039*Psi2)
4420        Chg(K)=-K*Ads(K)
4421        Sumchg=Sumchg+Chg(K)
4422        Sumads=Sumads+Ads(K)
4423        Next K
4424      Sigma2=Maxads*Sumchg/(1+Sumads)
4425      Gammaa=Maxads*Sumads/(1+Sumads)


4427      Rem  NOW CALC POTNL & CHARGE IN PLANE 3
4430      Psi3=Psi2-(Sigma1+Sigma2)*96.487/Gac
4440      Concat=Concelec : If Van>Vcat Then Concat=Concelec*Van
4450      Concat=Concat*Actcoef^(Vcat^2)
4460      Concan=Concelec : If Vcat>Van Then Concan=Concelec*Vcat
4470      Concan=Concan*Actcoef^(Van^2)
4480      Anionterm=Kan*Concan*Exp(0.039*Van*Psi3)
4490      Cationterm=Kcat*Concat*Exp(-0.039*Vcat*Psi3)
4500      Gammaan=Ns*Van*Anionterm/(1+Anionterm+Cationterm)
4510      Gammacat=Ns*Vcat*Cationterm/(1+Anionterm+Cationterm)


4515      Rem POTNL & CHARGE IN DIFFUSE LAYER
4520      Psi4=Psi3-(Sigma1+Sigma2+Gammacat-Gammaan)*96.487/Gcd
4530      If Vcat=Van Then 4532 Else 4540
4532      T5=Vcat*.0195*Psi4
4533       If Abs(T5)>87 Then Sigma4=-Sgn(Psi4)*1e35 : Goto 4580
4536      T6=.5*(Exp(T5)-Exp(-T5))
4538      Sigma4=-1.22*Sqr(Concelec)*T6
4539      Goto 4580
```

```
4540    X1=0.039*Psi4*Vcat : X2=0.039*Psi4*Van
4545     If Abs(X1)>87 Then Sigma4=-Sgn(Psi4)*1e35 : Goto 4580
4550    W3=(Exp(-X1)-1)/Vcat : W4=(Exp(X2)-1)/Van
4560    W5=Sqr(W3+W4) : If Psi4>0 Then W5=-W5
4570    Sigma4=0.61*Sqr(Concelec)*W5




4575    Rem COMPARE THE BALANCE OF CHARGE & IF GOOD ENOUGH, QUIT
4580    Estsigma1=-Sigma2-Gammacat+Gammaan-Sigma4
4590    Difference=Sigma1-Estsigma1
4600    If Abs(Difference)<1E-03 Then Goto 4850

4610    Rem THE NEXT STATEMENTS CAN BE ACTIVATED TO CHECK WHETHER
4611    Rem EQUATIONS ARE CONVERGING
4620    Rem     Print  Using"CURRENT PSIS ####.########";Psis(J);
4621    Rem     Print  Using" DIFF ###.#######";Difference

4630    D2=Sgn(Difference)*D1
4640    If Abs(Difference)>5 Goto 4770
4645    Rem WHEN DIFFERENCE LARGE, USE SUCCESSIVE INCREMENTS




4650    Rem WHEN DIFFERENCE IS SMALL ENOUGH USE LINEAR EXTRAPOLATION
4660    If Icount=1 Then Prepsi=Psis(J) : Goto 4720
4670    Slope=(Difference-Prediff)/(Prepsi-Psis(J))
4680    Intercept=Difference+Slope*Psis(J)
4690    Prepsi=Psis(J)
4700    If Slope=0 Then Psis(J)=Prepsi*0.999 : Goto 4720
4710    Psis(J)=Intercept/Slope
4720    Prediff=Difference
4730    Icount=Icount+1
4735    If Icount>30 Then Print "FAILED TO CONVERGE AT J= ",J:F=3*Fbest:Return
4740    If Icount=2 Then Psis(J)=Psis(J)+D2
4750    Goto 4180
4760    Rem   END OF LINEAR EXTRAPOLATION


4765    Rem ALTERNATIVE INCREMENTAL SEARCH
4770    If Icount=1 Then Icount=2
4772    Ii=Ii+1
4775    If Ii>50 Then Print "FAILED TO CONVERGE AT J=  ",J: F=3*Fbest:Return
4780    If Ii=1 Then 4810
4790    Prod=Difference*Prediff
4800    If Prod<0 Then D1=0.49*D1
4810    Prepsi=Psis(J) : Prediff=Difference
4820    Psis(J)=Psis(J)+D2
```

```
4830      Goto 4180
4840      Rem END OF CYCLE



4845      Rem   ALLOCATE THE APPROPRIATE VARIABLE TO BE THE PREDICTED VARIABLE.
4846      Rem   IN SOME CASES THIS MIGHT BE THE CHARGE, &/OR THE ADSORPTION.
4847      Rem   IN THIS CASE IT WAS THE FINAL SOLN CONC.
4848      Rem   IF ADSORPTION WERE TO BE PREDICTED FROM FINAL CONC, THEN
4850      Rem   Ye(J)=Gammaa
4855      Ye(J)=Concads-Suma*Wt*Area
4860      Estchg(J)=Sigma1+Sigma2



4870      Rem ACTIVATE THE NEXT STATEMENT TO CHECK ON THE PROGRESSIVE FIT
4871      Rem TO THE DATA ON THE SCREEN
4880      Rem    Print  J,Y(J),Ye(J),X(1,J),X(2,J),Estchg(J)



4885      Rem   THE NEXT STATEMENTS SPECIFY THE FUNCTION TO BE MINIMISED .
4886      Rem   IN THIS CASE, IT IS THE SQ OF THE DIFFERENCE BETWEEN THE
4887      Rem   LOGS OF OBSERVED & PREDICTED CONC.
4888      Rem   IF, FOR EXAMPLE, OBSERVED CHARGE WERE ALSO TO BE
4889      Rem FITTED, THE FUNCTION TO BE MINIMISED WOULD BE ALTERED.
4890      Disc=(Log(Y(J))-Log(Ye(J)))^2:  Rem  LOGS USED. ALTER IF REQUIRED
4895      F=F+Disc
4900      Next J


4910    Nevals=Nevals+1
4920    Print F,Time$
4930    Return



4940    Print "ERROR NUMBER    "Err,:Print "    AT LINE NUMBER    ",Erl
4950    End
```

Chapter B6

Applying models to soil

The models considered so far in Part B have been "models of models". That is, the reactions have been with adsorbants, such as goethite, that have been used as models of soil constituents. However soils are a good deal more complex. They obviously contain more than one constituent, and those constituents are far from pure. Further, the behaviour is different in several respects from that of pure substances. The effects of pH, though broadly similar, differ in detail; the effects of period of reaction are more marked; and the effects of concentration, as indicated by the shape of sorption graphs, are different. These differences are described in detail in Chapter A7 and the models that have been used to describe them are given in Chapter A8.

This Chapter presents four of the models. Two of these enable the modeller or vary the values of the parameters and to observe the effects on the output. Of these two, THETA is the simpler. It is appropriate when sorption has been measured at only one time. DIFPLUS is more complex in that the effects through time are included. It includes the effects of time on both the initial adsorption reaction and on the subsequent diffusion reaction. This program may also be used to explore the modelled effects of varying the concentration at chosen times in order to induce desorption or further sorption. Another avenue for exploration is the effects of varying the temperature. The modeller should, however, be careful to keep the model close to reasonable conditions in which, perhaps, the soil may have been incubated at one temperature and the null-point concentration measured at another temperature. Both of these programs are provided with a subroutine (ALPHA) to permit the calculation of the proportions of the ions present. This construction permits the easy substitution of alternative subroutines — such as for example ones derived from MOLLY or ZNOHP.

For both THETA and DIFPLUS some "suggested" values of the parameters are given. These are not the "correct" values. They should be used merely as starting points from which to explore the effects of varying the parameters. The results of such an exploration are given in Chapter A8. Armed with such experience, the modeller should be able to tackle the other two programs FITTHETA and FITRATE. Both these programs may be used to fit the model to specific sets of data. As presented, they are set up for some typical applications but the modeller may wish to make some of the "fixed" parameters variable and vice versa. This is fairly easily accomplished by varying the allocation of the parameters in subroutine EQUATION.

Programs DIFPLUS and especially FITRATE involve much iteration. They can therefore be slow. For example, each evaluation of a set of parameters using FITRATE with 15 data points (as indicated in the program) took nearly five minutes on an Olivetti M24 using GWBASIC. It would take longer on a computer with longer cycle time. Hence a run taking, say, 60 differing combinations of the parameters would involve five hours on the Olivetti. Increasing the frequency of sampling would increase this appreciably. Such long runs can, of course, be done overnight and so the lapsed time may be no greater than that involved in queueing for a large computer. However, when run using the Trump card (see Introduction) the program ran seven times faster. This means that parameters may be fitted over more convenient periods. It also means that programs such as DIFPLUS, which involve interaction with the operator, return the answer in a much shorter period and are therefore more convenient to use.

```
1    Rem                        THETA.BAS


10   Print"PRGM INTEGRATES A SERIES OF EQUATIONS TO TRY TO SIMULATE SORPTION"
20   Print"BY SOIL. IT IS ASSUMED THAT SOIL CONSISTS OF A SUITE OF SITES "
30   Print"WHICH DIFFER IN THE VALUE OF THE POTENTIAL - K1 IS CONST."
40   Print"THE END POINT IS LOCATED USING LINEAR EXTRAPOLATION."
45   Print"SETS UP A NORMAL DISTN. CENTERED ON XBAR & WITH ST DEV OF SIGMA"
50   Print"CLASS WIDTH =SIGMA/3"
55   Print "PRGM GENERATES VALUES FOR SORPTION OF P "
60   Print "AT GIVEN VALUES FOR CONCENTRATION. BY VARYING THE CONC.,"
64   Print " SORPTION CURVES CAN BE GENERATED AT GIVEN VALUES"
65   Print "OF THE PARAMS. SOME SUGGESTED VALUES OF THE PARAMS ARE GIVEN "
66   Print "(IN BRACKETS) TO FACILLITATE A FIRST RUN. THEN VARY THE VALUES!"
67   Print:Print


68   Dim Psi(30),Fraction(30)
69   Sumfract=0:Swtch=0


70   Input"MID POINT OF NORMAL DISTN. IN M.VOLTS   (-70) ",Xbar
72   Input"STANDARD DEVIATION OF DISTN. IN M.VOLTS (50)    ",Sigma
75   Nfractions=30
76   Input "DO YOU WANT TO SEE PRINTS OF VALUES FOR EACH SEGMENT(Y/N)",ANS$


77     Rem SET UP THE INITIAL DISTRIBUTION
78     For Ind=1 To Nfractions
80     Midpoint=Xbar-5*Sigma-Sigma/6+Ind*Sigma/3
85     Pi=3.141592653
90     Term=1/(Sigma*Sqr(2*Pi))
95     Prob=Term*Exp(-0.5*((Midpoint-Xbar)/Sigma)^2)
100    Fraction(Ind)=Prob*Sigma/3
105    Psi(Ind)=Midpoint
110    Sumfract=Sumfract+Fraction(Ind)
115    If Ans$="Y" Or Ans$="y" Then 116 Else 120
116    Print Ind;
117    Print Using" ####.#####";Psi(Ind),Prob,Fraction(Ind),Sumfract
120    Next Ind
121  If Ans$="Y" Or Ans$="y" Then 122 Else 123
122    Print "SEGMENT PSI     PROBABILITY     FRACTION    CUM.FRACTION"


123    Rem  INPUT THE OTHER PARAMETERS & VARIABLES
```

```
125   If Swtch =1 Goto 155: Rem "Swtch" PERMITS YOU TO CHANGE VALUES LATER
130   Input"VALUE FOR BINDING CONSTANT ON MICROMOLAR SCALE (10)     ",K1
133   Input "DO YOU WISH TO ENTER THE VALUE OF ALPHA ?     ",Answ$
134   If Answ$="Y" Or Answ$="y" Then 140 Else Gosub 5000: Goto 150
140   Input"PROPN OF THE P IONS PRESENT AS DIVALENTS (.1)    ",Alpha
150   Input"VALUE OF M IN MILLIVOLTS (145)       ",M
152   Input"MAXIMUM ADSORPTION IN MICRO G/G (3000)   ",Maxads
155   Input"CONC OF PHOSPHATE IN PPM   ",Cppm
156   C=Cppm*1000/31
162   Sumtheta=0


163   Rem BEGIN CALCULATIONS FOR EACH SEGMENT
165   For I= 1 To Nfractions
166   Psiest=Psi(I)
168     Cycle=1
169     Delta=1 : Inc=50 : Count=0


170        While Abs(Delta)>0.1
172        Inc2=Inc*Sgn(Delta)
175        If Psiest<-1800 Then Theta=0 : Goto 202
180        Kstar=K1*C*Alpha*Exp(2.0*Psiest*0.039)
200        Theta=Kstar/(1+Kstar)
202        Psinow=Psi(I)—M*Theta
204        Delta=Psinow—Psiest
205        If Abs(Delta)>100 And Cycle<3 Goto 235
206        If Cycle=1 Then  Prepsiest=Psiest: Psiest=Psiest+Inc2: Goto 228
214          Slope=(Predelta—Delta)/(Prepsiest—Psiest)
216          Intercept=Delta—Slope*Psiest
218          Prepsiest=Psiest
220          Psiest=—Intercept/Slope
228        Cycle=Cycle+1
230        Predelta=Delta
232        Goto 245


235        Count=Count+1 : Cycle=2
236        If Count=1 Goto 240
237        Prod=Sgn(Delta)*Sgn(Predelta)
238        If Prod<0 Then Inc=—0.49*Inc
240        Prepsiest=Psiest : Psiest=Psiest—Inc
242        Predelta=Delta
245        Wend


250     Sumtheta=Sumtheta+Theta*Fraction(I)
254     Psinow=Psi(I)—M*Theta
```

```
255    If Ans$="Y" Or Ans$="y" Then 256 Else 260
256       Print Using" ####.####  ####.####   ####.####";Psi(I),Theta,Psinow;
257       Print Using"  #.######     #.#####";Theta*Fraction(I),Sumtheta
260    Next I
261    Rem END CALCULATIONS


265    If Ans$="Y" Or Ans$="y" Then 266 Else 270
266    Print "   ORIG PSI    THETA    PSInow    THETA*FRACT    CUM.AMT "
267    Print "THETA IS THE FRACTION OF THE SEGMENT OCCUPIED BY P,"
268    Print "MUTIPLYING BY FRACT. GIVES THE WEIGHTED AMOUNT & "
269    Print "CUM.AMT GIVES THE PROGRESSIVE TOTAL"
270    Print Using"NET ADSORTION AS A % OF THE MAX IS ###.###";Sumtheta*100
280    Print:Print Using "THAT IS ####.#### MICROGM/G ";Sumtheta*Maxads:Print
300    Print "INPUT 1 TO CHANGE PROPS OF DISTN."
310    Print "      2 TO CHANGE SOIL CONSTANTS. "
320    Print "      3 TO CHANGE CONCENTRATION. "
325    Print "      4 TO QUIT. "
330    Input Swtch
340    On Swtch Goto 67,130,155,400
400    End




5000   Rem    SUBROUTINE  ALPHA
5001   Rem   CALCULATES THE PROPORTION OF ANIONS PRESENT AS GIVEN SPECIES

5005   Kdiss(1)=7.58E-03 : Kdiss(2)=6.166E-08 : Kdiss(3)=2.137E-13

5010   Input "IS ADSORBATE PHOSPHATE (Y/N)? ",Phos$
5015    If Phos$="Y" Or Phos$="y" Then 5200
5020   Print"DISS CONSTS FOR ADSORBING ACID. ENTER ZEROES FOR 2ND &/OR "
5030   Print"3RD CONST IF THEY DO NOT EXIST. NOTE  CONSTANTS - NOT LOG CONST"
5050   Print"FIRST",Tab(55); : Input Kdiss(1)
5070   Print"SECOND",Tab(55); : Input Kdiss(2)
5090   Print"THIRD",Tab(55); : Input Kdiss(3)


5200   Print"VALENCY OF ELECTROLYTE CATION",Tab(55); : Input Vcat
5220   Print"VALENCY OF ELECTROLYTE ANION",Tab(55); : Input Van
5235   Print"ELECTROLYTE CONC.(MOLAR SCALE)",Tab(55); : Input Concelec
5260   Print"pH",Tab(55); : Input pH
5275   ConcH=10^(-pH)


5300   Rem SIMPLE CALCULATION OF ACTIVITY COEFF FOLLOWS. ALTER IF YOU WANT
5310   Rem  MORE COMPLEX VERSION. THIS SIMPLE VERSION WILL NOT GIVE GOOD
5320   Rem VALUES FOR HIGH ELECTROLYTE CONC.
```

```
5330 Mu=0.5*(Van*Concelec*Vcat^2+Vcat*Concelec*Van^2)
5340 Actcoef=10^(-0.5*Sqr(Mu)/(1.0+Sqr(Mu)))
5350 Rem IF USING THERMODYNAMIC DISS CONSTANTS, ADJUST FOR IONIC STRENGTH.

5360 Kdiss1=Kdiss(1)/Actcoef^2 : Kdiss2=Kdiss(2)/Actcoef^4
5370 Kdiss3=Kdiss(3)/Actcoef^6
5400 T1=Conch^3+Conch^2*Kdiss1+Conch*Kdiss1*Kdiss2+Kdiss1*Kdiss2*Kdiss3
5410 Alf(1)=Kdiss1*Conch^2/T1
5420 Alf(2)=Conch*Kdiss1*Kdiss2/T1
5430 Alf(3)=Kdiss1*Kdiss2*Kdiss3/T1


5500 Alpha=Alf(2)
5510 Print "ALPHA  IS    ",Alpha


5600 Return
```

```
1    Rem                      DIFPLUS.BAS



10   Print "PROGRAM GENERATES SORPTION BY SOIL THRU TIME FOR GIVEN VALUES"
11   Print "OF THE PARAMETERS. SOME SUGGESTED VALUES ARE GIVEN IN BRACKETS"
12   Print "TO HELP ON THE FIRST RUN."
13   Print "THE IDEA IS TO TRY DIFFERENT PARAMETERS TO SEE THE EFFECT & SO"
14   Print "BECOME FAMILIAR WITH THE MODEL."
15   Print "REMEMBER, USE SMALL STEPS THROUGH TIME TO KEEP MODEL REALISTIC."



30   Rem PGME SETS UP A NORMAL DISTRIBUTION OF VALUES OF ELECTRIC POTENTIAL
40   Rem (PSI).IT THEN FINDS THE AMOUNT OF "ADSORBED" P AT SPECIFIED
50   Rem TIMES. NOTE, WHEN CONS IS CONST, THE PROCESS CAN BE TREATED AS A
55   Rem REVERSIBLE FIRST ORDER EQN. WITH THE FORWARD CONST INCLUDING CONC.
60   Rem IT IS THEN ASSUMED THAT  DIFFUSION BEGINS INTO THE SURFACE
70   Rem WITH THE RATE OF TRANSFER PROPORTIONAL TO THE SURFACE ACTIVITY.
80   Rem SURFACE CONC IS EXPRESSED AS THETA.
110  Rem NOTE AT ANY TIME THE SOLUTION CONC CAN BE CHANGED AND DESORPTION
120  Rem OR FURTHER ADSORPTION INITIATED



180  Dim Psi(30),Fraction(30),Psinow(30),Theta(30,20),Coefdiff(20),Time(20),Sumprod(2
190  Sumfract=0
192  Lprint "MODEL OF SORPTION THRU TIME WITH EACH STEP AT CONST SOLN CONC."
193  Lprint "INPUT VALUES ARE"
200  Rem SET UP NORMAL DISTN. CENTERED ON XBAR & WITH ST DEV OF SIGMA
210  Rem CLASS WIDTH =SIGMA/3



220  Input"MID POINT OF NORMAL DISTN (-70)   ",Xbar
230  Input"STANDARD DEVIATION (50)      ",Sigma
235  Lprint"MID POINT OF NORMAL DISTN  ",Xbar
236  Lprint"STANDARD DEVN              ",Sigma
237  Input "DO YOU WANT TO SEE OUTPUTS FOR ALL OF THE SEGMENTS? (Y/N)",Outpt$



239  Rem THE NEXT SECTION SETS UP THE INITIAL STATUS OF THE SEGMENTS
240  Nfractions=30
250   For Ind=1 To Nfractions
260   Midpoint=Xbar-5*Sigma-Sigma/6+Ind*Sigma/3
```

161

```
270     Pi=3.141592653
280     Term=1/(Sigma*Sqr(2*Pi))
290     Prob=Term*Exp(-0.5*((Midpoint-Xbar)/Sigma)^2)
300     Fraction(Ind)=Prob*Sigma/3
310     Psi(Ind)=Midpoint
320     Psinow(Ind)=Midpoint
330     Sumfract=Sumfract+Fraction(Ind)
340     If Outpt$="Y" or Outpt$="y" Then 343 Else 350
343      Print Ind;
345      Print Using" ####.#####";Midpoint,Prob,Fraction(Ind),Sumfract
350     Next Ind
351     If Outpt$="Y" Or Outpt$="y" Then 352 Else 353
352      Print"        MIDPOINT  PROBABILITY  FRACTION   SUMFRACTION"
353      Print :Print


355     Rem FURTHER INPUTS
360     Print"BINDING CONSTANT FOR ADSORBING IONS (eg DIVALENT P)",
361     Input" ON MICROMOLAR SCALE (10) ",Kb
362     Lprint "BINDING CONSTANT              "Kb


365     Print"DO YOU WISH TO ENTER DIRECTLY THE PROPN OF THE ADSORBATE"
366     Input "PRESENT AS THE ADSORBING ION AT THIS pH (ALPHA) (Y/N)",Ans$
367     If Ans$="Y" Or Ans$="y" Then 370 Else Gosub 5000 : Goto 375
370     Input"VALUE OF ALPHA  (eg .1) ",Alpha
375     Lprint"VALUE OF ALPHA                "Alpha


380     Input"VALUE OF M  IN MILLIVOLTS (145)   ",M
390     Input"VALUE OF M2  IN MILLIVOLTS (25)   ",M2
395     Lprint" M                             ",M
396     Lprint " M2                           ",M2


410     Input"MAXIMUM ADSORPTION IN MICRO G/G (3000)",Maxads
420     Input"SOLID STATE DIFFUSION CONST AT 25 DEG C   -PER DAY (1) ",Kdiff25
425     Input"ACTIVATION ENERGY FOR DIFF COEFF IN UNITS OF kJ/Mole  (80) ",En
426     Lprint" MAX. ADSORPTION  ",Maxads
427     Lprint" DIFFUSION CONST  ",Kdiff25
428     Lprint" ACTIVATION ENERGY  ",En


430     Energy=En*1000/8.3144
431     Input"VALUE FOR ALPHA-ARROW-FORWARD     (FOR P, 2) ",AlpharrowF
432     Input"VALUE FOR ALPHA-ARROW-BACKWARD    (FOR P, 0) ",AlpharrowB
433     Input"RATE CONSTANT FOR BACKREACTION FOR ADS  -PER DAY (10)    ",K2
434     Lprint "ALPHA ARROW FORWARD          ",AlpharrowF
```

```
435    Lprint"ALPHA ARROW BACK                ",AlpharrowB
436    Lprint"RATE CONST FOR BACK REACTION   ",K2


439    Input"MOLECULAR WT FOR ADSORBATE eg FOR P, 31  ",Molwt
440    Input"CONC OF ADSORBATE IN PPM      ",Conc
441    Conc=Conc*1000/Molwt
451      For I=1 To 30
453      Theta(I,0)=0
455      Next I
690    Input"TIME LAPSED IN DAYS —OR FRACTIONS OF DAYS (eg at first 0.0001) ",T
692    Time(1)=T
693    Input"TEMP IN DEG C    ",Ctemp : Abstemp=Ctemp+273



694    Rem  "HOUSEKEEPING" STEPS FOR FIRST CYCLE
696    Kbind=Exp(298*Log(Kb*1000000.0)/Abstemp)*1E-06
697    Kdiff=Kdiff25*Exp(Energy/298-Energy/Abstemp)
698    Coefdiff(1)=Kdiff
700    Tindex=1
705    Sumprod(1)=Time(1)*Kdiff
710    Sumads=0 : Sumfix=0
720    If Outpt$="Y" Or Outpt$="y" Then 730 Else 740
730    Print"COLUMNS ARE: PSI(0),THETA ADSORBD, THETA FIXED, PSI(now),THETA";
731    Print" ADS*FRACTION, SUM ADS, THETA FIX*FRACTION, SUMFIX"



735     Rem   MAIN LOOP STARTS
7         For I=1 To 30
7         If Tindex=1 Then  Psiest=Psi(I) Else  Psiest=Psinow(I)
750         Inc=10:Ii=0



760      Rem ITERATION CYCLE TO GIVE VALUES FOR PSI AT THIS TIME
770       For Cycle=1 To 50
785       Surfact=Kbind*Alpha*Conc*Exp(AlpharrowF*Psiest*0.039*298/Abstemp)
787       If Kbind*K2*Time(Tindex)>10 Then 790 Else 791
790       Theta(I,Tindex)=Surfact/(1+Surfact):Goto 809
791       K1star=K2*Surfact
792       K2star=K2*Exp(-AlpharrowB*0.039*Psiest*298/Abstemp)
793       Expterm=(K1star+K2star)*Time(Tindex)
794       If Expterm>80 Then Rateterm=1 : Goto 800
796       Rateterm=1-Exp(-Expterm)
800       Thetalast=Theta(I,Tindex-1)
801       Topterm=K1star*(1-Thetalast)-K2star*Thetalast
802       Botterm=K1star+K2star
```

163

```
803      Frontterm=Topterm/Botterm
805      Thetainc=Frontterm*Rateterm
806      Theta(I,Tindex)=Thetalast+Thetainc
809      Thetanow=Theta(I,Tindex)
820      Sumterm=0
830       For Jj=1 To Tindex
835       Factor=1-Theta(I,Tindex) : If Factor<=0 Then Factor=1E-03
840       Sumterm=Sumterm+(Theta(I,Jj)-Theta(I,Jj-1))*Sqr(Sumprod(Jj)/Factor)
850       Next Jj
860      Fixed=Sumterm*1.12838:REM 2/ROOT PI
910      Newpsiest=Psi(I)-(M*Thetanow+M2*Fixed)/(298/Abstemp)
920      Delta=Newpsiest-Psiest
930      Incpsi=Sgn(Delta)*Inc
940      If Abs(Delta)<0.1 Then  Cycle=50 : Goto 1050
945      If Abs(Delta)>1000 Goto 1031
950      If Cycle=1 Then Previouspsiest=Psiest:Psiest=Psiest+Incpsi:Goto 1025
990       Slope=(Previousdelta-Delta)/(Previouspsiest-Psiest)
1000       Intercept=Delta-Slope*Psiest
1010       Previouspsiest=Psiest
1020       Psiest=-Intercept/Slope
1025      Previousdelta=Delta
1030      Goto 1050
1031      Ii=Ii+1
1032      If Ii=1 Then 1040
1035      Prod=Delta*Previousdelta
1037      If Prod<0 Then Inc=0.49*Inc
1040      Previouspsiest=Psiest : Psiest=Psiest+Incpsi
1045      Previousdelta=Delta
1050      Next Cycle
1060    Rem END IF ITERATION CYCLE


1070     Psinow(I)=Psiest
1080     Sumads=Sumads+Thetanow*Fraction(I)
1090     Sumfix=Sumfix+Fixed*Fraction(I)


1093     Rem  PRINT OUT THE RESULTS FOR THE INDIVIDUAL SEGMENTS IF REQUIRED
1095      If Outpt$="Y" or Outpt$="y" Then 1100 Else 1130
1100     Print Using" ####.####    ###.####    ####.#### ";Psi(I),Thetanow,Fixed;
1110     Print Using" ####.####    ##.####   ";Psinow(I),Thetanow*Fraction(I);
1111     Print Using"####.#### ";Sumads;
1120     Print Using" ####.####    ##.#### ";Fixed*Fraction(I),Sumfix
1130      Next I
1135      If Outpt$="Y" or Outpt$="y" Then 1136 Else 1142
1136    Print " PSI(INIT)    THETA       PENETRTD   PSI(NOW)    THETA*FRAC";
1137    Print "   SUMADS  PEN*FRAC   SUM PENTRTD"
1140    Rem END OF MAIN LOOP
```

164

```
1142    Rem  PRINT OUT THE NET RESULT ON THE SCREEN & ON THE PRINTER
1145    Print
1150    Print Using"ADSORPTION IS NOW  ####.##### ADSORBED ";Sumads*Maxads;
1152    Print Using"PLUS  ####.###### FIXED  ";Sumfix*Maxads
1153    Totls=Maxads*(Sumfix+Sumads)
1154    Print Using"GIVING A TOTAL OF ####.###### Microgm/g  ";Totls
1156    Print Using"AT TIME ######.######     ";T
1158    Print Using"AND AT A CONC OF####.#### PPM";Conc*Molwt/1000


1160    Lprint
1165    Lprint Using"ADSORPTION IS NOW  ####.##### ADSORBED ";Sumads*Maxads;
1170    Lprint Using"PLUS. ####.###### FIXED  ";Sumfix*Maxads
1175    Lprint Using"GIVING A TOTAL OF ####.###### Microgm/g  ";Totls
1180    Lprint Using"AT TIME ######.######     ";T
1185    Lprint Using"AT A TEMPERATURE OF ###.## Deg  C";Ctemp
1190    Lprint Using"AND AT A CONC OF####.#### PPM";Conc*Molwt/1000


1191    Print
1193    Print "TO CONTINUE THRU TIME ENTER 1, TO RESTART WITH NEW CONC ENTER 2"
1194    Input "TO QUIT ENTER 3          ",Answ
1195    On Answ Goto 1197,440,1400


1197    Rem CONTINUE THROUGH TIME
1198    Tlast=T
1200    Input" TOTAL TIME NOW LAPSED  ",T
1201    If T<=Tlast Then Print "NEGATIVE TIME INCREMENT! RE-ENTER ": Goto 1200
1202    Input"HAS CONC OR TEMPERATURE CHANGED (Y/N)      ",C$
1203    If C$="Y" or C$="y" Then 1204 Else Goto 1208
1204    Input"NEW CONC IN PPM     ",Conc : Conc=Conc*1000/Molwt
1207    Input"TEMPERATURE IS NOW     ",Ctemp : Abstemp=Ctemp+273
1208    Kbind=Exp(298*Log(Kb*1000000.0)/Abstemp)*1E-06
1209    Kdiff=Kdiff25*Exp(Energy/298-Energy/Abstemp)
1210    Tindex=Tindex+1


1213    Rem SET UP THE NEW SEQUENCE OF LAPSED PERIODS*DIFFUSION COEFF
1215    Time(Tindex)=T-Tlast
1220    Coefdiff(Tindex)=Kdiff
1230      For J=1 To Tindex : Sumprod(J)=0 : Next J
1235      For J=1 To Tindex
1240        For Jj=J To Tindex
```

```
1260        Sumprod(J)=Sumprod(J)+Time(Jj)*Coefdiff(Jj)
1265        Next Jj
1270      Next J
1300    Goto 710



1400    End



5000  Rem   SUBROUTINE  ALPHA
5001  Rem  CALCULATES THE PROPORTION OF ANIONS PRESENT AS GIVEN SPECIES



5005  Kdiss(1)=7.58E-03 : Kdiss(2)=6.166E-08 : Kdiss(3)=2.137E-13


5010  Input "IS ADSORBATE PHOSPHATE (Y/N)? ",Phos$:If Phos$="Y" Or Phos$="y" Then 52C
5020  Print"DISS CONSTS FOR ADSORBING ACID. ENTER ZEROES FOR 2ND &/OR "
5030  Print"3RD CONST IF THEY DO NOT EXIST. NOTE  CONSTANTS - NOT LOG CONST"
5040  Lprint"DISS CONSTS FOR ADSORBING ACID. "
5050  Print"FIRST",Tab(55); : Input Kdiss(1)
5060  Lprint"FIRST",Tab(55),Kdiss(1)
5070  Print"SECOND",Tab(55); : Input Kdiss(2)
5080  Lprint"SECOND",Tab(55),Kdiss(2)
5090  Print"THIRD",Tab(55); : Input Kdiss(3)
5100  Lprint"THIRD",Tab(55), Kdiss(3)



5200  Print"VALENCY OF ELECTROLYTE CATION",Tab(55); : Input Vcat
5210  Lprint:Lprint"VALENCY OF ELECTROLYTE CATION  ",Vcat
5220  Print"VALENCY OF ELECTROLYTE ANION",Tab(55); : Input Van
5230  Lprint"VALENCY OF ELECTROLYTE ANION   ",Van
5235  Print"ELECTROLYTE CONC.(MOLAR SCALE)",Tab(55); : Input Concelec
5240  Lprint"ELECTROLYTE CONC.(MOLAR SCALE) ", Concelec
5260  Print"pH",Tab(55); : Input pH
5270  Lprint"pH                     ",pH
5275  ConcH=10^(-pH)



5300  Rem SIMPLE CALCULATION OF ACTIVITY COEFF FOLLOWS. ALTER IF YOU WANT
5310  Rem  MORE COMPLEX VERSION. THIS SIMPLE VERSION WILL NOT GIVE GOOD
5320  Rem VALUES FOR HIGH ELECTROLYTE CONC.
```

```
5330  Mu=0.5*(Van*Concelec*Vcat^2+Vcat*Concelec*Van^2)
5340  Actcoef=10^(-0.5*Sqr(Mu)/(1.0+Sqr(Mu)))
5350  Rem IF USING THERMODYNAMIC DISS CONSTANTS, ADJUST FOR IONIC STRENGTH.

5360  Kdiss1=Kdiss(1)/Actcoef^2 : Kdiss2=Kdiss(2)/Actcoef^4
5370  Kdiss3=Kdiss(3)/Actcoef^6
5400  T1=Conch^3+Conch^2*Kdiss1+Conch*Kdiss1*Kdiss2+Kdiss1*Kdiss2*Kdiss3
5410  Alf(1)=Kdiss1*Conch^2/T1
5420  Alf(2)=Conch*Kdiss1*Kdiss2/T1
5430  Alf(3)=Kdiss1*Kdiss2*Kdiss3/T1


5500  Alpha=Alf(2)
5510  Print "ALPHA IS      ",Alpha


5600 Return
```

```
10    Print"THIS PGM MAY BE USED TO EXPLORE THE PROCESS OF FITTING THE "
20    Print"MODEL TO SOIL DATA. SUPPOSE,FOR EXAMPLE, SORPTION WAS "
30    Print"DESCRIBED BY S=100*(Conc)^0.4. FIVE PAIRS OF VALUES FOR"
40    Print"SORPTION & CONC ARE:40,0.1; 75,0.5; 100,1; 190,5; & 251,10"
50    Print"IF YOU DON'T HAVE ANY DATA HANDY, TRY THESE. ENTER ALSO THE "
60    Print"SUGGESTED VALUES FOR THE OTHER PARAMETERS & WATCH THE FITTING "
70    Print"PROCESS. WARNING - THESE ARE NOT 'CORRECT' VALUES."
80    Print"USE THETA TO SEE THE EFFECT OF CHANGING THE PARAMETERS."


100   Dim Y(12),Conc(12),PredictY(12),Label$(30)
120   Dim Fraction(30),Psi(30)


125   Lprint "PROGRAM TO FIT SORPTION MODEL TO A SET OF SOIL DATA ": Lprint


130   Input"LABEL FOR DATA SET (eg TEST) ",Label$
135   Lprint : Lprint Label$ : Lprint


140   Input"VALUE FOR THE BINDING CONSTANT (10)              ",Kbind
141   Lprint"VALUE FOR THE BINDING CONSTANT                  ",Kbind


145   Input "MOLECULAR WT OF SORBATE eg FOR P 31        ",Molwt
146   Lprint"MOLECULAR WT OF SORBATE eg FOR P 31        ",Molwt


150   Input "VALENCY OF SORBING ION INCLUDING SIGN eg FOR P -2 ",Valency
151   Lprint"VALENCY OF SORBING ION INCLUDING SIGN eg FOR P -2 ",Valency


160   Rem IF DESIRED SUBROUTINE ALPHA COULD BE INSERTED HERE
170   Input"APRROX PROPN PRESENT AS SORBING ION (eg 0.1)       ",Alpha
171   Lprint"APRROX PROPN PRESENT AS SORBING ION                ",Alpha


175   Input"VALUE OF FEED-BACK COEF M   (150)              ",M
176   Lprint"VALUE OF FEED-BACK COEF M                  ",M


180   Input"VALUE FOR MAX ADSORPTION   (3000)             ",Maxads
181   Lprint"VALUE FOR MAX ADSORPTION                   ",Maxads


185   Input"NUMBER OF OBSERVATIONS              ",Nobs
190   Print"INPUT SORPTION, CONC"
195   Lprint : Lprint "  ENTERED VALUES FOR SORPTION & CONC "
200     For N=1 To Nobs
```

168

```
210      Input Y(N),Conc(N)
215      Lprint Y(N),Conc(N)
220      Next N


230   Input"NUMBER OF PARAMS TO BE VARIED -HERE TWO               ",Nparam
231   Lprint : Lprint


240   Input"FIRST EST OF THE MID-POINT OF THE DIST OF PSI'S(-70) ",Param(1)
241   Lprint"FIRST EST OF THE MID-POINT OF THE DIST OF PSI'S      ",Param(1)


270   Input"FIRST EST OF THE ST. DEV. OF THE DIST OF PSI'S(40)  ",Param(2)
271   Lprint"FIRST EST OF THE ST. DEV. OF THE DIST OF PSI'S      ",Param(2)



300   Rem TRANSFER TO SIMPLEX
310   Nmax=300 : Tolrance=0.005:Rem  THESE VALUES MAY BE VARIED
320   Print "VALUES FOR SORPTION & FITTED SORPTION WILL BE PRINTED"
321   Print "FOLLOWED BY THE PARAMS USED & THE RESIDUAL SUMS OF SQUARES"
330   Gosub 2000 : Rem Simplex


440   Print Using"JOB COMPLETED AFTER #### CYCLES";Nevals
450   Lprint : Lprint"FITTED VALUES FOR PARAMETERS " : Lprint
460     For Ind=1 To Nparam
470     Lprint Using"##  ########.#####";Ind,Param(Ind)
480     Next Ind
490   Print : Print
500   Rem RETURN TO EQUATION TO EVAL WITH BEST VALUES
510   Gosub 4000 : Rem Equation


540   Lprint : Lprint" FITTED POINTS    " : Lprint
550   Lprint"         CONC.        ADSORTION     FITTED ADS"
560   Sum=0 : Sumsq=0
570     For Ind=1 To Nobs
580     Lprint Using"   #####.#####";Conc(Ind),Y(Ind),PredictY(Ind)
590     Sum=Sum+Y(Ind) : Sumsq=Sumsq+Y(Ind)^2
600     Next Ind
610   Lprint : Lprint"VALUE OF F   ",F
620   Tssq=Sumsq-Sum^2/Nobs
630   Lprint : Lprint"TOTAL SUMS OF SQUARES   ",Tssq
640   Lprint : Lprint" R SQUARED              ",1-F/Tssq
650   End


2000  Rem INSERT SIMPLEX HERE
```

```
4000    Rem EQUATION
4010    Nevals=Nevals+1
4030    F=0
4040    Sumfract=0


4050    Xbar=Param(1) : Sigma=Param(2)


4060    Rem SET UP NORMAL DISTN. CENTERED ON XBAR & WITH ST DEV OF SIGMA
4070    Rem CLASS WIDTH =SIGMA/3
4080    Nfractions=30
4090      For Ind=1 To Nfractions
4100      Midpoint=Xbar-5*Sigma-Sigma/6+Ind*Sigma/3
4110      Pi=3.141592653
4120      Term=1/(Sigma*Sqr(2*Pi))
4130      Prob=Term*Exp(-0.5*((Midpoint-Xbar)/Sigma)^2)
4140      Fraction(Ind)=Prob*Sigma/3
4150      Psi(Ind)=Midpoint
4160      Sumfract=Sumfract+Fraction(Ind)
4165      Rem ACTIVATE THE NEXT STATEMENTS TO SEE THE BEHAVIOUR OF THE SEGMENTS
4170      Rem Print Ind;
4171      Rem  Print Using" ####.#####";Midpoint,Prob,Fraction(Ind),Sumfract
4180      Next Ind


4185    Rem NOW APPLY THE DISTRIBUTION TO CALC SORPTION
4190    For J=1 To Nobs
4200    Sumtheta=0 : Sumfixed=0 : Inc=10 :Ii=0
4210      For Ind=1 To Nfractions
4220      Psiest=Psi(Ind)
4230      Cycle=1 : Delta=1

4235        Rem START TO SOLVE THE EQUATIONS
4240        While Abs(Delta)>0.1

4243        Rem CALC SORPTION
4245        Expterm=-0.039*Valency
4250        If Expterm<-70 Then Theta=0 : Goto 4300
4260        Cmicromole=Conc(J)*1000/Molwt
4270        Surfact=Kbind*Cmicromole*Alpha*Exp(Expterm*Psiest)
4280        Theta=Surfact/(1+Surfact)

4290        Rem ADJUST POTENTIAL
4300        Psinow=Psi(Ind)+Sgn(Valency)*M*Theta

4305        Rem RE-ESTIMATE POTENTIAL
4310        Delta=Psinow-Psiest
4315        Incpsi=Sgn(Delta)*Inc
4320        If Abs(Delta)<0.1 Then 4495
```

```
4323              Rem IF NOT CONVERGING, TRY INCREMENTAL STEPS
4325              If Abs(Delta)>1000 Goto 4380
4330              If Cycle=1 Then Prepsiest=Psiest: Psiest=Psiest+Incpsi:Goto 4355
4335                Slope=(Predelta-Delta)/(Prepsiest-Psiest)
4340                Intercept=Delta-Slope*Psiest
4345                Prepsiest=Psiest
4350                Psiest=-Intercept/Slope
4355              Cycle=Cycle+1
4360              Predelta=Delta
4365              Goto 4495

4380              Rem  INCREMENTAL STEPS
4385              Ii=Ii+1 : If Ii=1 Then 4410
4390              Prod=Predelta*Delta
4400              If Prod<0 Then Inc=0.49*Inc
4410              Prepsiest=Psiest : Psiest=Psiest+Incpsi
4420              Predelta=Delta
4495              Wend
4500            Sumtheta=Sumtheta+Theta*Fraction(Ind)
4505           Next Ind
4510        PredictY(J)=Maxads*Sumtheta
4513        Rem NEXT STRATEMENT SHOWS PROGRESSIVE FIT. INACTIVATE IF NOT WANTED
4515        Print J,Y(J),PredictY(J)
4520        F=F+(Log(Y(J))-Log(PredictY(J)))^2
4530        Next J

4870     Print Using"RUN NO. ####";Nevals;
4880        For Ind=1 To Nparam
4890        Print Using"#####.####";Param(Ind);
4910        Next Ind
4920     Print Using"    F=   ####.######";F

4960     Return
```

```
1      Rem                    FITRATE.BAS


10     Print"THIS PGME SETS UP A DISTRIBUTION OF VALUES FOR PSI. IT THEN"
15     Print"USES THAT DISTRIBUTION TO CALCULATE ADSORBED & FIXED SORBATE"
20     Print"THRU TIMES OF .1,1,10,100,& 1000 DAYS AT 3 READ-IN CONCTRTNS."
25     Print"-THAT IS FOR 15 COMBINATIONS OF TIME & CONCENTRATION"
30     Print"MORE FREQUENT INTERVALS WOULD BE BETTER AND SHOULD BE USED"
35     Print"WITH A FAST COMPUTER. HOWEVER PRACTICAL RUN-TIMES"
40     Print"RESTRICT THE NUMBER OF PERIODS USING A PC."
45     Print : Print


50     Print"THE CALCULATED VALUES ARE COMPARED WITH READ-IN VALUES &"
55     Print"PARAMETERS ARE ADJUSTED TO FIND THE BEST MATCH."
60     Print"IT IS ASSUMED THAT ALPHA-ARROW FOR THE FORWARD REACTION IS "
65     Print"EQUAL TO THE VALENCY & THAT FOR THE BACK REACTION IS ZERO "
70     Print : Print


75     Print"AS SET UP, VALUES ARE ASSIGNED TO: MEAN PSI, M, & MAX. ADSORPTN"
80     Print"AND FITTED TO: THE BINDING CONSTANT,THE ST DEV OF PSI, M2, "
85     Print"THE DIFFUSION TERM, & THE RATE CONST FOR THE BACK REACTION"
90     Print"THESE CAN BE CHANGED BY CHANGING THE ALLOCATION OF THE PARAMS"
95     Print"(STATEMENTS 4070,4080), & BY CHANGES TO STATEMENTS 130 - 145"



100    Dim Y(15),Conc(3),Time(12),PredictY(15),Theta(30,8)
105    Dim Fraction(30),Psi(30),Psinow(30)



110    Lprint "PROGRAM TO FIT RATE EQUATIONS TO A SET OF SOIL DATA":Lprint
115    Lprint " VALUES READ IN ": LPRINT


120     Print : Print"ENTER VALUES FOR ASSIGNED PARAMETERS" : Print


130    Input"MIDPOINT OF DISTRIBUTION OF PSI (eg -90)         ",Xbar
131    Lprint"MIDPOINT OF DISTRIBUTION OF PSI                 ",Xbar


135    Input"MOLECULAR WEIGHT OF SORBATE eg FOR P, 31         ", Molwt
136    Lprint"MOLECULAR WEIGHT OF SORBATE eg FOR P, 31          ", Molwt


137    Input"VALENCY OF SORBING ION INCLUDING SIGN eg FOR P, -2  ", Valency
138    Lprint"VALENCY OF SORBING ION INCLUDING SIGN eg FOR P, -2 ", Valency


139    Rem SUBROUTINE ALPHA COULD BE ADDED TO PERMIT CALC OF ALPHA
```

```
140    Input"APRROX PROPN PRESENT AS SORBING ION (eg 0.1)          ",Alpha
141    Lprint"APRROX PROPN PRESENT AS SORBING ION                  ",Alpha


142    Input"VALUE OF FEED-BACK COEFICIENT  M  (eg 150)            ",M
143    Lprint"VALUE OF FEED-BACK COEFICIENT  M                     ",M


144    Input"VALUE FOR MAX ADSORPTION (eg 3000)                    ",Maxads
145    Lprint"VALUE FOR MAX ADSORPTION                             ",Maxads



146    Input"NUMBER OF OBSERVATIONS (HERE 15)                      ",Nobs


147    Print"INPUT OBSERVED VALUES FOR SORPTION."
148    Print"FIRST THE 5 OBS AT CONC 1 (EACH FOLLOWED BY A RETURN)"
149    Print"THEN THE 5 OBS AT CONC 2 ETC"
151      For N=1 To Nobs
152      Print "SORPTION ",N;:Input Y(N)
153      Next N
154    Print"INPUT THE 3 CONCENTRATIONS OF SORBATE (ppm)   "
155      For N=1 To 3 : Print "CONC ",N;:Input Conc(N) : Next N



156    Lprint "ENTERED OBSERVATIONS" : Lprint
157    J=0 : For I=1 To 3 : For K=1 To 5
158    J=J+1 : Lprint J,Conc(I),Y(J)
159    Next K : Next I



160    Input"NUMBER OF PARAMETERS TO BE VARIED - HERE 5            ",Nparam
161    Print :Print"GO AHEAD & GUESS THE STARTING VALUES OF THE PARAMS"
162    Print"THE OUTPUT WILL TELL YOU IF YOU ARE WAY OFF &, & IF SO"
163    Print"THEN YOU CAN COME BACK & TRY AGAIN" : Print


165    Input"FIRST EST OF THE BINDING CONSTANT                     ",Param(1)
166    Lprint"FIRST EST OF THE BINDING CONSTANT                    ",Param(1)


170    Input"FIRST EST OF THE  ST. DEV. OF THE DIST OF PSI         ",Param(2)
171    Lprint"FIRST EST OF THE  ST. DEV. OF THE DIST OF PSI        ",Param(2)


175    Input"EST OF SECOND FEED-BACK TERM M2                       ",Param(3)
176    Lprint"EST OF SECOND FEED-BACK TERM M2                      ",Param(3)


180    Input"EST OF THE DIFFUSION COEFF TERM                       ",Param(4)
181    Lprint"EST OF THE DIFFUSION COEFF TERM                      ",Param(4)


185    Input"EST OF RATE CONS FOR BACK REACTION OF ADS STEP        ",Param(5)
186    Lprint"EST OF RATE CONS FOR BACK REACTION OF ADS STEP       ",Param(5)
```

```
187    Print : Print
188    Print"VALUES FOR TIME, OBSERVED SORPTION, PREDICTED SORPTION, &"
189    Print"THE CONTRIBUTION TO THE SUMS OF SQUARES WILL BE PRINTED."
190    Print"THESE WILL BE FOLLOWED BY THE CURRENT VALUES OF THE"
191    Print"PARAMETERS & THE TOTAL SUMS OF SQUARES"


196    Rem TRANSFER TO SIMPLEX
197    Nmax=300
198    Tolrnce=0.005 : Rem ADJUST THESE NUMBERS AS REQUIRED
199    Switch=0
200    Gosub 2000 : Rem Simplex
210    Rem RETURN TO EQUATION TO INSERT BEST FIT VALUES
215    Switch=1
220    Print Using"JOB COMPLETED AFTER #### CYCLES";Nevals
230    Lprint : Lprint Using"JOB COMPLETED AFTER #### CYCLES";Nevals
235    Lprint : Lprint "FITTED VALUES ": Lprint
240    Lprint" NO.    TIME       CONC.         SORTION      FITTED SORP     ERR"
250    Gosub 4000


320    Lprint : Lprint"VALUE OF PARAMETERS "
330      For Ind=1 To Nparam
340      Lprint Using"###      ########.######";Ind,Param(Ind)
350      Next Ind
360    Lprint : Lprint
376    Sum=0 : Sumsq=0
380      For Ind=1 To Nobs
391      Sum=Sum+Log(Y(Ind)) : Sumsq=Sumsq+Log(Y(Ind))^2.0
400      Next Ind
405    Lprint"VALUE OF F    ",F
406    Tssq=Sumsq-Sum^2.0/Nobs
407    Lprint"TOTAL SUMS OF SQUARES    ",Tssq
408    Lprint" R SQUARED               ",1-F/Tssq
410    End

2000    Rem   TRANSFER   SIMPLEX  TO HERE



4000 Rem  EQUATION

4010    Rem INITIAL HOUSEKEEPING
4020    Nevals=Nevals+1
4040    If Param(4)<0 Then F=2*F : Goto 4810 : Rem DISCOURAGE IMPOSSIBLE VALUES
4050    F=0
4060    Sumfract=0
```

```
4065    Rem ALLOCATE PARAMETERS
4070    Kbind=Param(1) : Sigma=Param(2) : M2=Param(3) : Kdiff=Param(4)
4080    K2=Param(5)


4090    Rem SET UP NORMAL DISTN. CENTERED ON XBAR & WITH ST DEV OF SIGMA
4100    Rem CLASS WIDTH =SIGMA/3
4110    Nfractions=30
4120      For Ind=1 To Nfractions
4130      Midpoint=Xbar-5*Sigma-Sigma/6+Ind*Sigma/3
4140      Pi=3.141592653
4150      Term=1/(Sigma*Sqr(2*Pi))
4160      Prob=Term*Exp(-0.5*((Midpoint-Xbar)/Sigma)^2.0)
4170      Fraction(Ind)=Prob*Sigma/3
4180      Psi(Ind)=Midpoint
4190      Psinow(Ind)=Midpoint
4200      Sumfract=Sumfract+Fraction(Ind)
4210      Next Ind


4215    Rem START CALCULATIONS FOR EACH CONC & FOR EACH TIME
4220    J=0
4230      For Cindex=1 To 3
4240      Cmicromole=Conc(Cindex)*1000/Molwt
4250        For Ind=I To Nfractions : Theta(Ind,0)=0 : Next Ind
4255        Inc=10:Ii=0
4257        Rem THE NEXT SECTION SETS UP A SEQUENCE OF TIMES
4258        Rem DATA-READ STATEMENTS COULD ALSO BE USED.
4260        For Tindex=1 To 6
4270        T=10^(Tindex-3.0)
4280        Time(Tindex)=T
4290        Sumtheta=0 : Sumfixed=0


4296          Rem SOLVE THE EQUATIONS FOR EACH OF THE SEGMENTS
4300          For Ind=1 To Nfractions
4310          Psiest=Psinow(Ind)
4320          Cycle=1
4325          Delta=1


4329            Rem BEGINNING OF SECTION THAT SOLVES THE EQUATIONS
4330            While Abs(Delta)>0.1

4334            Rem   CALCULATE ADSORPTION
4335            Expterm=-0.039*Valency
```

175

```
4340        If Expterm<-70 Then Theta(Ind,Tindex)=0 : Fixed=0 : Goto 4510
4350        Surfact=Kbind*Cmicromole*Alpha*Exp(Expterm*Psiest)
4360        K1star=K2*Surfact
4370        Expterm=(K1star+K2)*Time(Tindex)
4380        If Expterm>70 Then  Rateterm=1 Else  Rateterm=1-Exp(-Expterm)
4400        Thetalast=Theta(Ind,Tindex-1)
4410        Topterm=K1star*(1-Thetalast)-K2*Thetalast : Botterm=K1star+K2
4420        Frontterm=Topterm/Botterm : Thetainc=Frontterm*Rateterm
4430        Theta(Ind,Tindex)=Thetalast+Thetainc
4440        Thetanow=Theta(Ind,Tindex)

4445        Rem CALCULATE PENETRATION
4450        Sumterm=0
4460          For JJ=1 To Tindex
4470          Factor=1-Theta(Ind,Tindex) : If Factor<=0 Then Factor=1E-03
4475          Mult=Sqr((T-Time(JJ-1))*Kdiff/Factor)
4480          Sumterm=Sumterm+(Theta(Ind,JJ)-Theta(Ind,JJ-1))*Mult
4490          Next JJ
4500        Fixed=Sumterm*1.12838   :Rem  2/ROOT PI

4505        Rem  ADJUST POTENTIAL
4510        Psinow(Ind)=Psi(Ind)+Sgn(Valency)*(M*Thetanow+M2*Fixed)

4515        Rem RE-ESTIMATE THE POTENTIAL
4520        Delta=Psinow(Ind)-Psiest
4525        Incpsi=Sgn(Delta)*Inc
4530        If Abs(Delta)<0.1 Then  4680


4533        Rem IF NOT CONVERGING, TRY INCREMENTAL STEPS
4535        If Abs(Delta)>1000 Then 4640
4540        If Cycle=1 Then Prepsiest=Psiest:Psiest=Psiest+Incpsi :Goto 4630
4580          Slope=(Predelta-Delta)/(Prepsiest-Psiest)
4590          Intercept=Delta-Slope*Psiest
4600          Prepsiest=Psiest
4610          Psiest=-Intercept/Slope
4630        Cycle=Cycle+1
4635        Predelta=Delta
4637        Goto 4680


4639        Rem INCREMENTAL STEPS
4640        Ii=Ii+1 : If Ii=1 Then 4650
4645        Prod=Predelta*Delta
4647        If Prod<0 Then Inc=0.49*Inc
4650        Prepsiest=psiest : Psiest=Psiest+Incpsi
4660        Predelta=delta
4680    Wend
```

```
4681            Rem END OF SECTION THAT SOLVES THE EQUATIONS
4684            Rem CALCULATE THE AGGREGATES
4685            If Thetanow<1E-20 Then 4700
4690            Sumtheta=Sumtheta+Thetanow*Fraction(Ind)
4700            Sumfixed=Sumfixed+Fixed*Fraction(Ind)
4710            Next Ind
4715            Rem  END OF CALCULATION FOR THE SEGMENTS


4720            If T=0.01 Then 4780: Rem NO DATA AT THIS TIME
4730            J=J+1
4740            PredictY(J)=Maxads*(Sumfixed+Sumtheta)
4745            Rem THE NEXT STATEMENT PERMITS CHECKING OF THE FIT.
4746            Rem INACTIVATE IF NOT REQUIRED
4750            Print Using"###  ####.#      ########.###    ";J,T,Y(J);
4755            Print Using"########.###  ";PredictY(J);
4760            Discrep=(Log(Y(J))-Log(PredictY(J)))^2.0
4765            F=F+Discrep
4770            Print Using"  ####.####";Discrep


4771          Rem LIST TO PRINTER ON FINAL RETURN TO THIS SUB PROGRAM
4772          If Switch=0 Goto 4780
4774          Lprint Using"###  ####.#     ####.### " ;J,T,Cmicromole*Molwt/1000;
4775          Lprint Using" ########.##### ";Y(J),PredictY(J),Discrep
4780          Next Tindex
4790        Next Cindex


4800    Rem PRINT STATUS ON SCREEN AT END OF EACH SET OF PARAMS
4810    Print Using"RUN NO. ####";Nevals;
4820      For Ind=1 To Nparam
4830      Print Using"#######.####";Param(Ind);
4840      Next Ind
4850    Print Using"   F=#######.######";F
4855    If Nevals>1 Then 4865
4860    Input"CARE TO HAVE A WISER GUESS AT THE STARTING ESTIMATES(Y/N)",St$
4865    If St$="Y" Or St$="y" Goto 165
4870    Return
```

Chapter B7

Describing the reaction of ions with soil

In previous chapters we have been concerned with mechanistic models of the reaction of ions with soil. Thus we have been concerned with both description and explanation. In the present chapter we are concerned solely with description — that is with choosing functions that can be used to reduce many observations to a few numbers that adequately describe the behaviour. As will be shown, this has further relevance to this book in that it also involves solving simultaneous equations using iterative methods.

The first aspect to be considered is the basis on which we may judge whether a particular function is adequate. Consider an experiment in which a substance is permitted to react with soil. For convenience let us imagine that the substance is phosphate — thus simplifying the terminology (the argument, of course, applies for any reactant). Suppose that the experimenter has chosen a solution:soil ratio and an appropriate background electrolyte, and that he has mixed samples of a soil with a series of solutions of differing phosphate concentration. After a specified period he separates soil and solution and measures the phosphate concentration in the solution. From the change in concentration he calculates the phosphate sorbed by the soil (Ps). He then seeks an appropriate function to relate Ps to the solution concentration of phosphate at the end of the experiment. The question is, what criteria should he use to choose that function? The essence of the problem is that the experimenter has measured the concentration in solution and any errors in this measurement will be transferred to the calculation of sorption. To illustrate the problem, let us assume that the Freundlich equation is the true model — that is, all errors are errors of measurement rather than errors due to lack of fit of the model to the data.

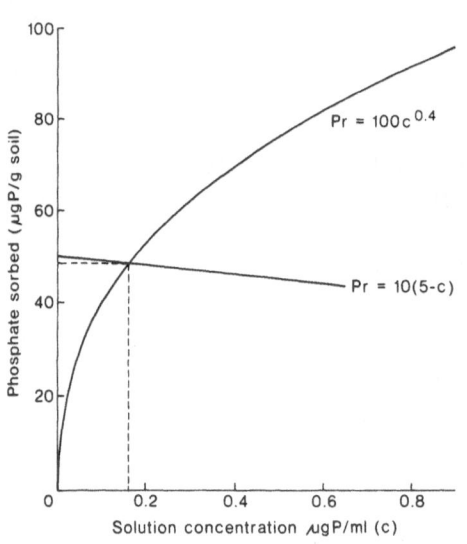

Fig. B7.1 A representation of equation B7.4 and B7.5. The intersection represents the simultaneous solution and is at the point: sorption = 48.4, concentration = 0.163.

Then

$$Ps = a(c + e)^b \tag{B7.1}$$

where c is the solution concentration of phosphate, e is the error of measurement and a and

178

b are parameters. Because the errors are made in analysing the solution concentrations of phosphate, it is reasonable to assume that the errors are proportional to the observations:

$$e = kc \qquad\qquad (B7.2)$$

Hence

$$Ps = a(1 + k)^b \, c^b \qquad\qquad (B7.3)$$

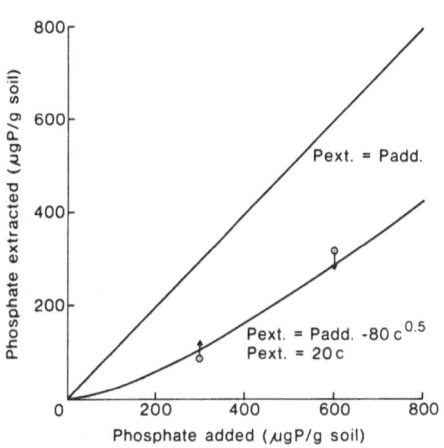

Fig. B7.2 A representation of equation B7.6. The curved line indicates the solution to B7.6 and B7.7 for different levels of addition to phosphate. The points represent supposed observed values and the tips of the arrows from these points indicates the value for phosphate extracted as claculated from Eq B7.6 alone — that is ignoring the simultaneous equation.

Because b is less than unity, the error in predicting Ps is much smaller than the error in observing c. For example, if b has the value of 0.4, a 10% error in c leads to only a 4% error in predicting Ps; a 30% error in c leads to only a 11% error in Ps. The equations chosen to illustrate this point give simple relations. Nevertheless it seems a reasonable general conclusion that calculation of the fit of such a function will always underestimate the error. It is tempting to speculate that the resulting good "fits" of functions to data is one reason for the popularity of this kind of research. Let us now consider alternative ways of handling this problem.

The alternatives can be best illustrated using a numerical example. Suppose that the true model for sorption is:

$$Ps = 100 \, c^{0.4} \qquad\qquad (B7.4)$$

Assume that the solution:soil ratio (Ss) is 10 and that the initial phosphorus concentration in solution was 5 μg P/ml. Then the values for c and for Ps which should be obtained can be calculated by solving B7.4 simultaneously with:

$$Ps = Ss(5 - c) \qquad\qquad (B7.5)$$

These two equations are illustrated in Fig.B7.1. Their solution can be obtained by iteration (using program PREDX — see later). It is Ps = 48.4, c = 0.163. Suppose, however, that the observer makes a 30% error in measuring c and records a value of 0.212. There are four possible approaches.

1) Simple application of equation B7.4 (the usual approach)
    Observed concentration   0.212
    "Observed" sorption (calculated from the change in concentration)   47.9
    Sorption predicted from observed concentration (from B7.4)   53.7
                                                                          error   11%

2) Attempt to predict the observed variable — concentration
    "Observed" sorption   47.9
    Concentration corresponding to observed sorption (from B7.4)   0.159
    Difference from observed concentration   (.212 — .159) .053
                                                                          error   32.5%

3) Base the calculation on the simultaneous equations that described the system B7.4 and B7.5 and predict sorption
    "Observed" sorption   47.9
    Predicted sorption   48.4
                                                                          error   1.1%

4) As 3) but predict the observed variable — concentration
    Observed concentration   0.212
    Predicted concentration   0.163
                                                                          error   30%

Of these alternatives, number 1, and especially number 3, give a false impression of the precision. Number 2 over-estimates the error slightly. However if the solution:soil ratio were much smaller, then errors in the measurement of the solution concentration would have little effect on the calculation of "observed" sorption. This would be true if the solution:soil ratio were 0.2:1 — that is, at approximately field moisture content. In this case, calculation of the expected concentration will be unaffected and the calculated error will be very close to the true error. When the solution soil ratio is large, however, only alternative 4) correctly reflects the error of measurement.

Before considering programs to deal with this problem, let us consider a further aspect. To illustrate it, let us suppose that the experimenter has added phosphate to soil and after a given period has extracted the soil with a 0.5M solution of sodium bicarbonate using a solution:soil ratio of 20:1. Suppose that the true equation to describe these results is:

$$P_{ext.} + P_{add.} - 80c^{0.5} \qquad\qquad\qquad\qquad\qquad\qquad \text{B7.6}$$

The following equation is also true:

$$P_{ext.} + 20c \qquad\qquad\qquad\qquad\qquad\qquad\qquad\qquad \text{B7.7}$$

The resulting relation between phosphate extracted and phosphate added is shown in Fig. B7.2. For a level of addition of phosphate of 300 μg P/g soil, the expected value for the concentration of phosphate in the extractant is 5.55 μg P/g. Suppose that, due to experimental error, the observed concentration is only 4.5 μg P/ml and hence the observed value for phosphate extracted is 90 μg P/g. If we simply substitute this observed value for concentration into equation B7.6, the value for phosphorus extracted predicted by the equation is 130. Thus the magnitude of the error is greatly exaggerated. Furthermore, the calculated line will wander from the true line — the greater the error in the observation, the further the calculated line will differ from the true line. This kind of problem arises whenever sorption is related to an extra variable and is plotted against that variable — in this case against phosphate added. Thus it also arises if sorption is related to both concentration and to time and is plotted against time.

Programs to deal with these problems follow. PREDX is a simple program to solve simultaneous equations like B7.4 and B7.5. It extends the equation by including terms for time and for amounts of phosphate already present in the soil. To use it in its simpler forms, set time equal to one and native phosphate equal to zero. FITFUNC shows how this approach can be used to fit a series of alternative functions to a set of sorption data. Note that the Gunary equation has been written in a fashion that permits first guesses of the parameters by analogy with the output of the Langmuir equation. In this form, the location of the end point is more efficient than in other forms. The equations have been written in a way that permits fairly easy modification — for example by adding a term for native phosphate or by adding terms for time. The equations can also be modified to deal with the problems involved in solving equations B7.6 and B7.7.

```
10    Print"PROGRAMME TO CALCULATE THE SOLUTION CONC OF P AS A
20    Print"RESULT OF ADSORPTION. IT SOLVES SIMULTANEOUS EQUATIONS AND
30    Print"THUS RELATES CONC TO EXTERNAL VARIABLES :  SS RATIO, P ADDED
40    Print"& TIME.
50    Print"A LINEAR EXTRAPOLATION IS USED TO MINIMISE THE DIFFERENCES
60    Print"BETWEEN THE TWO SIMULTANEOUS EQUATIONS.


80    Print"THE PROGRAMME NEEDS AN INITAL STARTING POINT. IT CAN PROVIDE
90    Print"IT FOR ITSELF OR YOU CAN PUT IN YOUR OWN ESTIMATE.
100   Print"STATEMENTS 610 TO 670 WILL RESCUE THE PROCEDURE IF THE ESTIMATE
110   Print"IS TOO FAR OUT.


130   Print"EQUATION ASSUMED TO BE DESCRIBING ADSORPTION IS:"
140   Print"          Psorbed= A*(T^m)*(C^n) — Q"
150   Rem PROGRAM COULD BE REWRITTEN TO USE ANY OTHER APPROPRIATE FUNCTION.
160   Print


170   Print"INPUT COEFFICIENTS"
180   Print
190   Input"LINEAR COEFFICIENT A               ",A
200   Input"EXPONENT FOR TIME m                ",M
210   Input"EXPONENT FOR CONCENTRATION n       ",N
220   Input"P PRESENT IN SOIL AT ZERO TIME Q   ",Q
230   Print


240   Print"INPUT EXTERNAL VARIABLES"
250   Print
260   Input"SOLUTION/SOIL RATIO                ",SsRatio
270   Input"INITIAL P CONC IN SOLUTION          ",Iconc : Padd=Iconc*SsRatio
280   Input"TIME OF SHAKING IN UNITS APPROPRIATE TO THE VALUE OF A  ",T


290   Input"SHALL I GUESS THE STARTING POINT FOR YOU — Y/N?",St$
300   If St$="y" Or St$="Y" Then  310  Else 340
310     EstConc=((Padd+Q)/A)^(1/N)*T^(—M/N)
320     Print"INITIAL ESTIMATE",EstConc
330     Goto 360
340     Input"YOUR ESTIMATE OF CONCENTRATION IN SOLUTION  ",EstConc
```

```
360    StartPoint=EstConc
370    I=1 : Difference=1



375      Rem START CYCLE
380      While Abs(Difference)>0.03
390      EstPAds=A*T^M*EstConc^N-Q


410      Rem CALCULATE THE SOLUTION SOIL RATIO THAT WOULD GIVE THIS PAIR OF
420      Rem VALUES FOR CONCENTRATION & ADSORPTION.
440      ApparentSSRatio=(Padd-EstPAds)/EstConc
450      Difference=ApparentSSRatio-SsRatio


460      Rem    FOR FIRST STEP, INCREASE CONCENTRATION BY 10%.
470      If I>1 Then   520
480        PreviousConc=EstConc
490        EstConc=1.1*EstConc


500      Rem FOR OTHER STEPS, USE LINEAR EXTRAPOLATION TO EST END POINT.
510      Goto 590
520      Slope=(Difference-PreviousDifference)/(PreviousConc-EstConc)
530      PreviousConc=EstConc
540      EstConc=EstConc+Difference/Slope


550      Rem IF RELATION IS CURVED EXTRAPOLATION TO IMPOSSIBLE NEG. VALUES
560      Rem MAY OCCUR.
570        If EstConc<0 Then EstConc=PreviousConc*0.2
590    PreviousDifference=Difference
600    I=I+1


610      Rem IF NOT MINIMISING, RESTART.
620      If I>5 And EstConc>1000*StartPoint Then   630 Else 680
630        EstConc=StartPoint*0.5
640        Print"re-estimate      ",EstConc
650        StartPoint=EstConc
660        Goto 370
680      Wend



690    Print
700    Print Using"VALUE FOR CONCENTRATION    ####.#####";EstConc
```

```
710    Print Using"VALUE FOR P ADSORBED         ####.#### ";EstPAds
720    Print Using"AT A SOLN. SOIL RATIO OF      ###.#";SsRatio
730    Print Using"FOR AN INITIAL CONC OF        ####.##";Iconc
740    Print Using"AFTER TIME                    ####.##";T
750    Print
760    Print" NO. OF ITERATIONS   ",I


770    Print"TO CHANGE THE PARAMETERS OF THE EQUATION, ENTER 1"
780    Print"TO CHANGE SS RATIO                      ENTER 2"
790    Print"TO CHANGE P CONC                        ENTER 3"
800    Print"TO CHANGE TIME                          ENTER 4"
810    Input"TO QUIT                                 ENTER 5    ",Index
820    On Index Goto 170,260,270,280,830
830    End
```

```
1    Rem                    FITFUN.BAS


5    Print"PGM TO FIT NON LINEAR EQUATIONS TO DATA FOR SOIL SORPTION"
10   Print"IT INVOLVES A SIMULTANEOUS FIT OF THE DESIGNATED EQUATION "
15   Print"AND THE EQUATION INVOLVING THE SS RATIO"
20   Print"INPUT IS FROM THE KEYBOARD. YOU WILL BE ASKED TO ENTER THE"
25   Print"OBSERVED SOLUTION CONCENTRATION & THE INITIAL CONCENTRATION"
30   Print"TIME IS NOT INCLUDED AS A VARIABLE BUT THE PGM CAN EASILY"
35   Print"BE MODIFIED TO INCLUDE IT"


50   Dim Y(50),Ye(50),X(4,50),Psorb(50),Param(20),Title$(40)


70   Input"HEADING FOR THIS DATA SET    ",Title$
75   Lprint Title$
80   Print"NO. OF POINTS     " : Input Nobs


85   Input"DID YOU USE THE SAME SOLN/SOIL RATIO FOR ALL POINTS (Y/N) ",Ssr$
87   Nx=2
90   If Ssr$="Y" Or Ssr$="y" Then Nx=1:Input "THE RATIO WAS   ",Ssr
93   For J=1 to Nobs : X(2,J)=Ssr : Next J
95   Rem INCREASE Nx IF EXTRA VARIABLES SUCH AS TIME ARE INCLUDED


100  Print"TO LOG TRANSF Y,ENTER 1" : Input Logtrans
105  If Logtrans=1 Then Lprint : Lprint "MINIMISE ON LOG OF CONC "


110  Print"ENTER OBSERVED CONC, INITIAL CONC, & (IF VARIED) S/S RATIO"
115  Lprint : Lprint "ENTERED VALUES FOR CONC, INITIAL CONC, & SS RATIO"
120    For I=1 To Nobs
125    Lprint
130    Gosub 600
140    Next I
145  Print
146  Lprint
150  Print"TYPE 1 TO PROCEED , 2 TO ALTER AN ENTRY" : Input C1
155  If C1=2 Then 156 Else 157
156  Print"WHICH ONE IS U.S.":Input I:Lprint"REPLACE ",I:Gosub 600:Goto 150

157  Print"ENTER 1 FOR FREUND, 2 FOR LANG, 3 FOR GUNARY, 4 FOR DOUBLE "
158  Input"LANG.,5 FOR TOTH     ",Switch
```

```
163    Lprint:Lprint
164    On Switch Gosub 166,167,168,169,170
165    Goto 175
166    Lprint "FREUNDLICH EQUATION" : Nparam=2 : Return
167    Lprint "LANGMUIR EQUATION  " : Nparam=2 : Return
168    Lprint "GUNARY EQUATION    " : Nparam=3 : Return
169    Lprint "DOUBLE LANGMUIR    " : Nparam=4 : Return
170    Lprint "TOTH  EQUATION     " : Nparam=3 : Return


175    Input "SHALL WE INCLUDE A TERM FOR NUTRIENT ALREADY THERE(Y/N)   ",Qq$
180    If Qq$="Y" Or Qq$="y" Then Nparam=Nparam+1


182    Input "IF YOU WANT TO MAKE YOUR OWN FIRST EST OF PARAMS, ENTER 1 ",Prm
183    If Prm=1 Then 185 Else 205
185    Print"FOR IDENTITY OF PARAMS SEE SUBROUTINE EQUATION"
190      For I=1 To Nparam
195      Print"PARAMETER",I; : Input"          ",Param(I)
200      Next I
201    Goto 230

202    Rem SOME STARTING PARAMETERS
205    On Switch Gosub 210,212,214,216,218: Goto 230
210    Param(1)=100 : Param(2)=.4 :Param(3)=10 : Return
212    Param(1)=1 : Param(2)=200 : Param(3)=10 : Return
214    Param(1)=1 : Param(2)=1000 : Param(3)=20 : Param(4)=10 : Return
216    Param(1)=10 : Param(2)=50 : Param(3)=.1 : Param(4)=500 : Param(5)=10
217    Return
218    Param(1)=1 : Param(2)=1000 : Param(3)=0.2 : Param(4)=10


230    Nmax=300 : Tolrnce=.001
231    Rem DECREASE THE VALUE FOR TOLRNCE IF PGM JUMPS OUT TOO SOON
232    Print"VALUES FOR PARAMETERS & FOR VALUE OF FUNCTION BEING MINIMISED"


235    Rem TRANSFER TO SIMPLEX
240    Gosub 2000


242    Rem RETURN TO EQUATION TO EVAL BEST FIT
245    Gosub 4000
250    Ysum=0 : Ysumsq=0 : Devsq=0
260      For I=1 To Nobs
265      Yi=Y(I) : Yei=Ye(I) :If Logtrans=1 Then Yi=Log(Y(I)) : Yei=Log(Ye(I))
270      Ysum=Ysum+Yi
280      Ysumsq=Ysumsq+Yi^2
```

186

```
290    Devsq=Devsq+(Yi-Yei)^2
300      Next I
310    Sumssqy=Ysumsq-Ysum^2/Nobs
330    R1=1.0-Devsq/Sumssqy
335    Print : Print"R SQ      ",R1
340    Print : Print
350    Lprint : Lprint"R SQ      ",R1
360    Lprint : Lprint


365    Rem PRINT OUT RESULTS
370    Print"TOTAL SUMS OF SQUARES",Sumssqy," RESID SUMS OF SQUARES",Devsq
375    Print : Print
380    Print"VALUES OF PARAMETERS"
385    Print
390      For I=1 To Nparam
395      Print Using"PARAMETER ##";I;
396      Print Using"    ######.#####";Param(I)
400      Next I
405    Lprint : Lprint
410    Lprint"TOTAL SUMS OF SQUARES",Sumssqy," RESID SUMS OF SQUARES",Devsq
415    Lprint : Lprint
420    Lprint"VALUES OF PARAMETERS"
425    Lprint
430      For I=1 To Nparam
435      Lprint Using"PARAMETER ##";I;
436      Lprint Using"    ######.#####";Param(I)
440      Next I
445    Lprint : Lprint
450    Lprint"PLOTTING POINTS"
460    Lprint : Lprint Title$ : Lprint
470    Lprint "  NO      Y       PREDICTED Y    DEV     X'S & EST SRPTN."
480      For I=1 To Nobs
490      Lprint Using"  ######.###";I,Y(I),Ye(I),Y(I)-Ye(I);
500        For J=1 To Nx
510        Lprint Using"  ######.###";X(J,I);
520        Next J
525      Lprint Using"  ######.###";Psorb(I)
540      Next I
550    Lprint
560    Lprint Using"NUMBER OF CYCLES  ####";Nevals
570    Lprint


575    Input "ANOTHER EQUATION ON THE SAME DATA (Y/N)?    ",Answ$
576    If Answ$="Y" Or Answ$="y" Then 162
580    Input "ANOTHER SET OF DATA (Y/N)?   ",Answ$
585    If Answ$="Y" Or Answ$="y" Then 70
```

```
590    End


599    Rem   SUBROUTINE TO READ IN DATA
600    Print I;"  OBS CONC"; : Input "        ",Y(I)
605     Lprint Using "#####.####";Y(I);
610      For J=1 To Nx
620      If J=1 Then Input "       INIT CONC      ",X(J,I)
625      If J=2 Then Input "       SS RATIO       ",X(J,I)
628      If J=3 Then Input "       TIME           ",X(J,I)
630      Next J
640    If Ssr$="Y" Or Ssr$="y" Then Nxx=Nx+1 Else Nxx=Nx
650    For J=1 To Nxx
660      Lprint Using "#####.####";X(J,I);
670    Next J
680    Return



2000   Rem   TRANSFER SUBROUTINE SIMPLEX TO HERE




4000   Rem        SUBROUTINE EQUATION


4005   Rem ALLOCATE PARAMETERS
4010   A=Param(1)
4020   B=Param(2)
4030   C=Param(3)
4040   D=Param(4)
4042   E=Param(5)
4045   Rem IS SOMETIMES DESIRABLE TO PREVENT SOME VALUES OF THE PARAMETERS eg
4050   If B<0 And Switch=3 Then F=2*Fbest : Goto 4830
4070   F=0.0


4080     Rem START THE CYCLE
4090     For Jj=1 To Nobs
4100     Ssratio=X(2,Jj)
4110     Padd=X(1,Jj)*Ssratio
4130     Estconc=Y(j)
4140     Startpoint=Estconc
4150     Index=1 : Count=0
4160     Adjust=0.5
```

```
4170       Difference=1

4195       Rem REPEAT THE NEXT SECTION UNTIL DIFFERENCE IS SMALL ENOUGH
4200       While Abs(Difference)>1E-03

4205       Rem  SELECT THE EQUATIOM
4210       On Switch Gosub 4230,4260,4290,4320,4370
4215       Rem  THESE EQUATIONS MAY BE MODIFIED AS DESIRED
4220       Goto 4430

4230    Rem  FREUNDLICH
4240       Estpads=A*Estconc^B
4242     Rem IF TIME WERE A VARIABLE THEN Estpads=A*Estconc^B*X(3,J)^C
4245       If Qq$="Y" Or Qq$="y" Then Estpads=Estpads-C
4247       Rem THE TERM FOR "NATIVE" P WOULD ALSO VARY WITH TIME
4250       Return

4260    Rem   LANGMUIR
4270       Estpads=A*B*Estconc/(1+A*Estconc)
4275       If Qq$="Y" Or Qq$="y" Then Estpads=Estpads-C
4280       Return

4290    Rem   GUNARY
4300       Estpads=A*B*Estconc/(1.0+A*Estconc+C*Sqr(Estconc))
4305       If Qq$="Y" Or Qq$="y" Then Estpads=Estpads-D
4310       Return

4320    Rem   DOUBLE LANGMUIR
4330       Term1=A*B*Estconc/(1+A*Estconc)
4340       Term2=C*D*Estconc/(1+C*Estconc)
4350       Estpads=Term1+Term2
4355       If Qq$="Y" Or Qq$="y" Then Estpads=Estpads-E
4360       Return

4370    Rem   TOTH
4372       Term1=A*B*Estconc : Term2=(1.0+(A*Estconc^C))^(1/C)
4380       Estpads=Term1/Term2
4385       If Qq$="Y" Or Qq$="y" Then Estpads=Estpads-D
4390       Return

4400       Rem CALCULATE THE SOLUTION SOIL RATIO THAT WOULD GIVE THIS PAIR OF
4410       Rem VALUES FOR CONCENTRATION & ADSORPTION.
4430       Apparentssratio=(Padd-Estpads)/Estconc
4450       Difference=Apparentssratio-Ssratio

4470       Rem   FOR FIRST STEP, INCREASE CONCENTRATION BY 1%.
```

```
4480        If Index=1 Then  4490 Else 4530
4490           Previousconc=Estconc
4500           Estconc=1.01*Estconc
4505           Goto 4620


4510           Rem FOR OTHER STEP, USE LINEAR EXTRAPOLATION TO EST END POINT.
4530           Slope=(Difference-Previousdifference)/(Previousconc-Estconc)
4540           Previousconc=Estconc

4545           Rem ACTIVATE THE NEXT STATEMENT TO WATCH PROGRESSIVE ESTIMATES
4550           Rem Print Jj,Index,Estconc,Estpads,Difference
4560           Estconc=Estconc+Difference/Slope

4580           Rem IF RELATION IS CURVED EXTRAPOLATION TO IMPOSSIBLE NEGATIVE
4590           Rem VALUES MAY OCCUR.
4600           If Estconc<0 Then Estconc=Previousconc*0.2
4620         Previousdifference=Difference
4630         Index=Index+1

4640         Rem IF NOT MINIMISING, RESTART.
4660         If Index>5 And Estconc>1000*Startpoint Then 4670 Else 4730
4670           Estconc=Startpoint*Adjust
4700           Adjust=0.5*Adjust
4710           Count=Count+1
4720           Index=1

4725         Rem  PROVIDE ESCAPES - THESE ARE NEEDED IF PARAMETERS GO TO VALUES
4726         Rem  FOR WHICH THE EQUATIONS DON'T CONVERGE
4730           If Index>30 And Abs(Difference)<1e-2 Then Difference=0
4740           If Index>50 Then Print "NOT MINIMISING":F=3*Fbest:Goto 4890
4745           If Count>20 Then Print "NOT MINIMISING ":F=3*Fbest:Goto 4890

4750         Wend
4755         Rem END OF THE SECTION IN WHICH SIM. SOLN TO EQUATIONS FOUND

4765       Rem ALLOCATE VALUES
4770       Ye(Jj)=Estconc
4780       Psorb(Jj)=Estpads
4805     Rem IF LOG TRANS REQUIRED MINIMISE ON LOG VALUES
4810       If Logtrans=1 Then 4815 Else 4817
4815       F=F+(Log(Y(Jj))-Log(Ye(Jj)))^2
4816       Goto 4820
4817       F=F+(Y(Jj)-Ye(Jj))^2
4820       Next Jj
4830     Nevals=Nevals+1
4840     Print Using"###";Nevals;
4850       For Ii=1 To Nparam
```

```
4860      Print Using"  #####.#########";Param(Ii);
4870      Next Ii
4880   Print Using" ######.#####";F
4890   Return
```